Planning the Greenspaces of Nineteenth-Century Paris

PLANNING THE GREENSPACES OF NINETEENTH-CENTURY PARIS

RICHARD S. HOPKINS

Louisiana State University Press   Baton Rouge

Published by Louisiana State University Press
Copyright © 2015 by Louisiana State University Press
All rights reserved
Manufactured in the United States of America
First printing

Frontispiece: Postcard, "A Full Load" (*Un Chargement complet*), ca. 1900. Children's carriage ride in the Parc des Buttes Chaumont (Nineteenth Arrondissement)

Designer: Barbara Neely Bourgoyne
Typeface: Calluna
Printer and binder: Maple Press

Library of Congress Cataloging-in-Publication Data

Hopkins, Richard S., 1961–
 Planning the greenspaces of nineteenth-century Paris / Richard S. Hopkins.
  pages cm
 Includes bibliographical references and index.
 ISBN 978-0-8071-5984-2 (cloth : alk. paper) — ISBN 978-0-8071-5985-9 (pdf) — ISBN 978-0-8071-5986-6 (epub) — ISBN 978-0-8071-5987-3 (mobi) 1. Urban parks—France—Paris—History—19th century. 2. Parks—France—Paris—History—19th century. 3. Public spaces—France—Paris—History—19th century. 4. Paris (France)—History—19th century. I. Title.
 SB484.F7H67 2015
 712'.50944361—dc23

2014035918

The paper in this book meets the guidelines for permanence and durability of the Committee on Production Guidelines for Book Longevity of the Council on Library Resources. ∞

**Library
University of Texas
at San Antonio**

CONTENTS

	PREFACE	ix
	Introduction	1
1.	A New Garden Capital *National Prestige and Municipal Efficiency*	14
2.	Public Health and the Greening of Paris	37
3.	Greenspace as a Workplace	60
4.	Just around the Corner *The Neighborhood and the Urban Park*	93
5.	Cultivating Broader Communities	130
	Conclusion	153
	APPENDIX: PARKS AND SQUARES	161
	NOTES	165
	BIBLIOGRAPHY	195
	INDEX	213

MAPS

•••• ••••

1. Square Louis XVI, ca. 1865   97
2. Parc des Buttes Chaumont, 1867–93   98
3. Square Parmentier, ca. 1875   101
4. Parc Montsouris, ca. 1875   104
5. Square Montholon, before 1874   113
6. Square Montholon, after 1874   114

PREFACE

•••• ••••

During my first trip to Paris, I was struck by the beauty and profusion of urban greenspace. A park, garden, or planted square seemed to lie just beyond each turn. The landscaping and arrangements appeared thoughtfully done and maintained with tremendous care. More important, these were animated spaces, full of people. It was summertime, and many tourists were visiting the larger, well-known parks such as the Tuileries and the Luxembourg Gardens. Parisians were there as well, but it was in the smaller, neighborhood squares that the city's residents outnumbered other park goers. There were mothers with children playing nearby; teenagers sitting on the grass with their friends, book bags strewn beside them, and smoking cigarettes; businessmen and women taking a midday break in the sunshine; men on the down-and-out with weathered faces insisting in loud voices on some point or another; affectionate couples oblivious to passersby; mature women strolling along the gravel path led by a sprightly little dog; and bronzed, rugged-looking men focused intently on a game of *pétanque*. I had visited parks in the United States that were equally as lively, but there was something different in Paris. It seemed to me that these Parisians inhabited these greenspaces with a particular zeal. They were living out life's most intimate exchanges between family, friends, and lovers in public and appeared to be wholly unconcerned about the gaze of others. Every person occupied the space boldly assured of his or her right to be there. They took ownership over the space. Many travelers to Paris comment on the abundance of parks and their popularity. I wanted to know more about these greenspaces. Why did Paris have so many parks and squares, and how had the special connection Parisians seemed to have to them come about? Was it always so? Thus, long before I ever considered research into Parisian greenspaces, the seeds of this project had already been planted.

As we will encounter the various Parisian forms of urban greenspace in the pages of this book, I will take a moment here to discuss briefly those terms, without suggesting any kind of rigid definitions. A *bois* (wood), the largest of these urban constructions, and exemplified by the Bois de Boulogne and the Bois de Vincennes, can be understood in the context of urban greenspace as an expanse of open, planted space for public use, constituted by a mix of wooded and grassy areas, several acres in size, and designed in a naturalistic style. The Bois de Boulogne and Bois de Vincennes are distinct from a general bois, meaning simply a stand of trees, and the much larger *forêt,* which is a more densely wooded space, rural, and in the nineteenth-century often the setting for a hunt. An urban *parc* (park), as in the Parc des Buttes Chaumont, Parc Montsouris, and Parc Monceau, is much smaller than a bois. It often contains a similar blend of small groupings of trees and stretches of lawn. The term *jardin* (garden) connotes a more structured greenspace designed in a fairly formal style. It may be small, or quite large as in the case of the Jardin du Luxembourg, Jardin des Tuileries, and Jardin des Plantes, and include some areas that are informally designed with trees and grass. The *square* (pronounced *skwar* in French), which so captured my imagination, is a small, public, neighborhood garden, generally surrounded by a fence, and often situated in the middle of a *place,* or an opening in the built environment at the intersection of several streets. Despite the term, squares can assume any shape. Moreover, not every place in Paris has a planted square, and a square can sometimes be called a place, as in the case of the Place des Vosges, for example. The French borrowed the word *square* from the English term for a small, enclosed, urban greenspace, and *square* came into common use in France along with an adaptation of the form during the mid-century park development program. The English had, in fact, borrowed the French term *équerre,* a tool for making right angles, centuries early to describe their own public spaces, generally square but not always green, which were becoming common in their cities, particularly London.[1] A *promenade,* in nineteenth-century Paris, was at once the activity of taking a stroll or a horse or carriage tour, and the greenspace within which one might do that.

A study of Parisian greenspaces such as this would have been impossible without the support, friendship, intellectual insight, and assistance of so many along the way. Institutional support was critical to the realization of this project. A fellowship, awarded by the International Dissertation Research Fellowship Program of the Social Science Research Council and funded by the Andrew W. Mellon Foundation, made possible the lengthy and detailed site research essential to this project. Grants from the Arizona State University Graduate College and Department of History also greatly facilitated my archival research and subsequent completion of this study.

Archivists and librarians in Paris extended a warm welcome to me and contributed their professional expertise. I am particularly grateful to Brigitte Lainé at the Archives de Paris, whose help was invaluable in locating innumerable rarely used documents and sources. Also, Anne Remond at the Bibliothèque de l'École du Breuil shared generously her knowledge of that institution's collections. Both offered expert guidance and kind encouragement over many months of research. Thanks also to the many workers at the Archives de Paris, Archives Nationales, Archives de la Préfecture de Police de Paris, Bibliothèque Historique de la Ville de Paris, and the Bibliothèque Nationale de France for their kind assistance.

Warm thanks goes to my friends and colleagues at Arizona State University: Victoria Thompson, Kent Wright, and Susan Gray, who read complete drafts of this work and offered their insightful and most valuable feedback. I am also indebted to others at ASU and beyond who commented on chapters, or offered their warm encouragement and welcome advice throughout this journey: Roger Adelson, Anna Chichopek-Gajraj, Chouki El Hamel, Martha Hanna, Katie Jarvis, Colin Jones, Brian Newsome, Laurie Manchester, Hava Samuelson, Chris Szuter, Patty Turning, and Mark Von Hagen. Most especially, I thank my dear friend and mentor, Rachel Fuchs. Her counsel and guidance at every stage of this project were essential in bringing it to fruition. She is a profoundly generous and inspiring scholar. My indebtedness to her is quite simply immeasurable.

At the Louisiana State University Press I have been most fortunate to have had Alisa Plant, with her expertise and enthusiasm, guide me through preparing this book for publication. I am very grateful, as well, to the anonymous readers for their intellectually stimulating and tremendously helpful suggestions. To the many friends and fellow historians of modern France in the United States and in Europe who shared their comments, questions, and camaraderie throughout the evolution of this project, I express my profound appreciation. My deepest gratitude goes to my family, who never wavered in support of my efforts to bring this book to life even though it meant long periods of absence while conducting research. Finally, I offer my warmest and most heartfelt thanks to Zac, Senna, and Bosie for their patience, confidence, and unconditional support.

Planning the Greenspaces of Nineteenth-Century Paris

INTRODUCTION

••• •••

The people evidently understand that they are at home; that it is for their especial behoof that the gardens have been constructed; they know that in pulling up a flower it is their own property they are destroying; and moreover they evince a respectful gratitude for the hands that have given them these pleasant places to resort. The establishment of the public squares in Paris is an eminently social idea.
—IRISH HORTICULTURALIST WILLIAM ROBINSON, 1869

••• •••

The wealth of public greenspace in Paris today and its place in peoples' lives originated in the mid-nineteenth century. Advocated by social reformers, spurred by national concerns, and instituted by a municipal government with broad powers, many of the parks and squares of Paris emerged during the Second Empire's massive renovation of the capital city, and continued to be constructed during the Third Republic. As emperor, Napoleon III had pressed for a renovation of the city, and his prefect of the Seine, Georges-Eugène Haussmann, administered the sweeping changes. State and municipal powers oversaw the park-building program and determined much of its character. Still, the interplay among Parisian citizens, and between city residents and the municipal park service, the *Service des Promenades et Plantations,* which was responsible for building and maintaining parks, helped also to shape Parisian urban greenspaces in significant ways. Together, over the second half of the nineteenth century, the park designers, city administrators, and the people of Paris who utilized the parks and squares engineered and reengineered the nature of public greenspace in the capital city, blurring the lines between creator and user. This dynamic then ensured the development and sustainability of a system of urban space that could address a multiplicity of desires and needs, from the state level to the individual, and one that would become relevant to the lives of ordinary citizens. By acknowledging the role of city residents in this way, this work challenges an understanding of mid-century urbanism as merely a top-down process. Parisian greenspace was, and continued to be, an urban form rooted in and shaped significantly by its public function. Thus, the evolution of Parisian public parks and squares demonstrates the way in which city residents at all social levels understood and addressed the challenges of rapid urbanization, exercising agency and fashioning their environment, and how those same elements transformed over time, reflecting social and cultural changes in France. Moreover, this examination of Parisian greenspace development, a program begun under the direction of Napoleon III's lead designer-engineer for parks, Jean-Charles Adolphe Alphand, and continued during the Third Republic, contributes to a nuanced, although non-apologetic, history of the Second Empire, pointing out some continuities with preceding and subsequent regimes.

Historians have written about the dramatic, physical transformation of Paris from a chaotic, closed medieval city to a rational, open, and modern city in the nineteenth century with its wider boulevards, regularized facades, increased commercial space, and resulting diminished affordable housing. Much of this discussion has focused on the goals of political authorities and changes to the urban infrastructure. With its emphasis on Napoleon III and his prefect Georges-Eugène Haussmann, David Pinkney's *Napoleon III and The Rebuilding of Paris* remains a foundational work that considered the political and economic history of the physical changes to the cityscape during the Second Empire.[1] Subsequent studies of Haussmann's life and career likewise contributed much to understanding the motivations and results of the mid-century alterations emphasizing thoroughfares, transportation and sewer systems, and housing.[2] Other scholars have approached the period through a wide array of social and cultural histories focused on the nature and production of urban space and social relations.[3] Throughout the literature on mid-century Paris, however, treatment of the parks and greenspaces of the city has remained fairly tangential, or has focused on them in the context of the history of landscape architecture, even though they were such an immediately identifiable (and now iconic) aspect of the redesign.[4] This book argues that the nineteenth-century greenspace development program in Paris was not simply urban embellishment, although it certainly served that purpose well; rather, its very nature and scope necessarily involved environmental management on a scale unprecedented in the French capital, affecting the urban landscape and lives of city residents. Placing greenspaces, as both conceptual and social spaces, in the forefront of a discussion of changes to the cityscape adds a new dimension to the topics of the mid-century transformation of Paris in particular and urban development in general.

The methodological framework of this study is rooted in an understanding of those public greenspaces as at once physical and experiential. It considers the issues surrounding the advocacy, creation, and management of each of the many parks, squares, and gardens in Paris—that is, the rationales and motivations behind their development through to their actual construction or modification. Indeed, reformers and doctors, and park builders and administrators, all believed a prosperous, modern city had to be a healthful city and emphasized quality of life for all citizens in their arguments for expansive public greenspace development. Then, in shifting focus to the park user as a participant in defining these city spaces, this work adds an additional layer to our understanding of the urban landscape absent in much of the discussion of Parisian urbanism. Through combining multiple perspectives of planners, social reformers, park workers, and, significantly, the neighborhood inhabitants who used the spaces, my approach considers parks and squares as at once conceived and engineered, *and* as spaces of human activity underscoring the actual use of the spaces by city residents. This methodology is informed by the work of geographical theorist Yi-Fu Tuan, who proposed bringing together the subjective and objective viewpoints in order to understand meaning in landscapes. Tuan asserted that meaning is extracted by first ordering reality from different perspectives, and then contemplating landscape as a habitat. Those perspectives are the vertical or functional view in which the observer is, in a sense, looking down on a two-dimensional plane, and the horizontal, or moral/aesthetic view, wherein the observer is within the landscape. To a certain extent, these correspond respectively to the planning and construction of parks and squares, and the experience of those spaces. According to Tuan, constructed landscapes are intended habitats; thus, to fully comprehend meaning, one must understand how that particular landscape relates to human need and experience.[5]

The records of the Service des Promenades et Plantations, the nineteenth-century Parisian park service charged with creating and maintaining urban greenspaces, were a rich archival resource for this book. They contained not only information about construction and management of greenspaces within Paris, but also significant detail concerning the public's use of the parks. Often, insight into everyday life and what was important to city residents appeared in unexpected ways, such as in documentation of new constructions or contained in numerous requests made to the service by park concessionaires or visitors. Citizen petitions, often attached to or discussed in administrative records with details about the petitioners and their claims, were particularly helpful in showing how local residents envisioned and used park space. Personnel files and guard reports revealed much about the complex relationship between park workers and the public, and they helped to fill in some of the gaps in nineteenth-century municipal police records concerning criminal activity and infractions of park rules.[6] Statutes, regulations, and contracts with concessionaires provided information concerning the place of commerce in the park, shedding light on the quotidian activity in greenspaces. Interdepartmental circulars, brochures, *ordres du jour* (executive directives communicated to park guards), and memoranda, which were never intended to be seen outside of the park service, reveal the inner workings and culture of that agency. Budgets, cost projections, maps, and plans were other valuable archival resources to explore the various design options that were under consideration and to confirm, in some cases, park constructions. Together, these records show that, although the park service was most certainly a central player in shaping public greenspace, it did not operate in isolation or in the absence of public participation.

In addition to the administrative records of the Service des Promenades et Plantations, tracts and treatises on housing, sanitation, and public health issues published by reformers, public hygienists, physicians, scientists, and landscape architect/engineers provide an understanding of the philosophy and rationale behind greenspace development. Newspaper accounts of the alterations to the city and activities in the park, and journals concerned with architecture, public works, and garden design contain much of the popular and professional discussion concerning greenspaces and the alterations to Paris. Finally, memoirs, written by Parisians and visiting foreigners, often described in lively ways not only the impression of Paris and greenspaces during the nineteenth century, but also what was happening in the parks and squares.

They are particularly useful in that they augment published and widely read guidebooks such as those by Adolphe Joanne and the Karl Baedeker and John Murray firms, which contain descriptions and histories of the many larger parks and gardens but often lack the kind of reportage one locates in travelers' accounts. These then complement administrative records and convey still more about the understanding and actual use of public greenspace.

This investigation into the rationale, establishment, management, and use of greenspace reveals several key themes. They do not appear in equal measure, and none dominates the others. Rather, taken together, they illuminate some of the process by which Parisians came to shape and reshape their world, or engineer their space, their use and their understanding of it. The first of these themes is the impact of positivism on greenspace development. This builds on historian Nicholas Papayanis's work, similarly linking intellectual currents and urban developments, and explores how, emerging as it did with roots in Saint-Simonian utopian socialism, social reform, science, and engineering, the park-building project was a decidedly positivist program.[7] Eschewing the dogmatic, repressive religion of humanity that positivist philosopher Auguste Comte espoused, the Service des Promenades et Plantations functioned much more in concert with the earlier ideas of Henri de Saint-Simon concerning the role of science and technology in understanding and improving the world, ideas that men such as Émile Littré, and even Napoleon III, had championed and made particularly popular during the Second Empire.[8] Alphand and others in the park service embraced the notion present in Comte's early work that human society and reasoning had evolved to an elevated stage in the nineteenth century and that its products—the sciences and technology—should be marshaled to ameliorate the conditions of modern life.[9] Theirs was a blend of study, science, progress, and social benefit. This emphasis is evident in many of the communications between the park service and park goers, and in the countless inquiries and studies that preceded its decisions. Composed largely of engineers seeking workable solutions and efficient systems to address the challenges of creating and maintaining constructed *natural* spaces in the *unnatural* urban environment, the park service was pragmatic to its core and fairly receptive toward public input if it could improve results.

A second theme in this book involves concepts of public and private. On a fundamental level, greenspace was public space. Not only did it exist outside of the private realm of the home, it was outside of private property and distinct from commercial and residential urban spaces, although both existed within the parks minimally in the form of concessions and employee residences. Questions arose concerning the rights and responsibilities of private citizens while visiting the public parks, access and use of this public space, and the protection and safety of private citizens within it. The increase of greenspace throughout the capital and in closer proximity than in the past to peoples' homes engendered a proprietary feeling over the space among city residents. With no legal or actual ownership rights, citizens and citizen groups were nevertheless keenly interested in the management, maintenance, and use of parks and squares they suddenly found close to their homes. Although they were constructed public spaces, parks and squares were designed, in many respects, to function as extensions of the home. With the physical health of children connected to greenspace and put forward as a rationale for its development, parks included recreation areas where parenting, child play, care, and discipline occurred in the open and before the public's gaze. Thus, the lines between the private space of the home and the public space of the city tended to blur in the parks, affecting both the residents' perception and use of the space and the administration's management of it. I contend that this blurring of the boundary between the private and the public influenced the way in which municipal authorities and city residents acted together to shape the urban environment and helped create public greenspaces that would be relevant to daily life in the city.

The third theme concerns questions of rights and liberty. A perception of fundamental rights to greenspace combined with an acknowledgment of the limitations to liberty in the exercise of those rights (a legacy of the discourse on rights during the French Revolution) informed the attitudes of the park builders as well as the residents of Paris. Although this sentiment was rarely explicitly articulated, the concept stood as one of the key rationales for providing greenspace, and it appeared in various ways in the correspondence and interactions between officials and city residents. Early nineteenth-century reformers and hygienists spoke of inherent rights when they considered the "right" to air and light and salubrious housing. Children, in particular, they claimed, had a "right" to a healthy environment. A tacit acknowledgment of rights of access and to use of greenspace permeates many of the decisions of the parks administration. The greatest limitation on rights within the park occurred at the precise point where the exercise of one's rights infringed upon the rights of another—an echo of the *Déclaration des droits de l'homme et du citoyen.* Thus, Parisian public greenspace can be read as an expression in the urban landscape of France's revolutionary past.

Finally, human agency, or the extent to which ordinary people made choices and took actions that helped shape their world, is a critical component of this book. This particular thematic element is evident in considerations of the spatial practices, or what Michel de Certeau called "ways of operating," of the park employees and the park visitors.[10] De Certeau argued that spatial practices can secretly reconfigure space and the conditions of social life; in essence, ways of operating, movement through the space, even if that is movement of thought, can change a place into a space. Certeau articulated this difference between a place and a space, identifying a place, or a *lieu,* as the visible and tangible, and a space, or an *espace,* as invisible, experiential, and achieved through movement. "Space," he claimed, "is a practiced place."[11] The force of spatial practices in shaping an understanding of urban greenspace is evident in the actions of workers taking a shortcut through the park on the way to or from work, or parents who left children to the care of others in a square, or gangs of youths appropriating a bench or statue for nightly gatherings, or elites parading in a display of wealth around the lakes in the Bois de Boulogne. These ways of operating and thinking about greenspace altered the way some Parisians from all corners of society understood the meaning and social function of the parks in their city.

This thematic focus also draws on the work of anthropologist Victor Turner. According to Turner, at times persons in societies may experience a moment of unification through a shared experience with strangers or acquaintances, a process he identified as "experiential *communitas.*"[12] The feeling of unity is so strong that the assembly or group identifies the experience as being shared and immediately seeks to organize its sense of itself as a group. This organizing impulse he called "normative *communitas.*" For Turner, *communitas* is a "modality of social relationship," or a means of community formation, rather than a location.[13] Turner further argued that the sense of camaraderie can be so strong and its positive impact on social organization so great that the group seeks to reconstitute the sensation through repetitions of the experience of liminality, rites of passage, or rituals of communitas.[14] Turner was primarily concerned with liminality, the secular and the sacred, social structure, and ritual as means of transcending social structures in societies. However, his concepts can be a helpful way of exploring how people, who may have had a loose or even no connection to each other outside of the greenspace, might form bonds and coalesce into a recognizable group, either by sharing the park space and an experience there, or by sharing interest in use of the space. The ostensibly unstructured nature of greenspace, and its position within yet set apart from the city, in some instances, helped to encourage communitas. At the very least, Parisian greenspace offered great opportunities for it, which were not always present in other urban spaces. Turner's concept of communitas highlights not only how greenspace could foster community, but how residents and park goers, through repetition of communal activities such as military concerts, skating afternoons, and mothers tending to their children, encouraged the Service des Promenades et Plantations to reconsider often the design and function of the parks. This exercise of agency, sometimes collective, is a significant aspect of the engineered natural spaces of nineteenth-century Paris and represents a level of public participation in urban planning even in the absence of any coordinated public outreach.

Concepts of community, identity, spatial practices, rights, liberty, and public and private realms were all are linked to the physical form and style of the mid-century Parisian park design, a landscape form that had a long history. Indeed, early modern greenspace construction and landscape design practices in France and in Europe laid the foundation for the design environment within which mid-nineteenth-century engineers worked. Landscape designers and park builders had long investigated the representational nature of design and how it might be used to speak to different visions of the nation, the individual, and the state. Indeed, the English style of garden design (so evident in the nineteenth-century redesign of Paris) constituted an eighteenth-century rejection of the formal French style epitomized in the work of André Le Nôtre.[15] Not surprisingly then, the English style of garden design came into vogue in France in the decades leading up to the French Revolution, sometimes as a celebration of a new social order, and sometimes as pure fashion.[16] Not until the mid-nineteenth century would the fluid, ostensibly natural, *jardin anglais* style reemerge as a rigorously engineered, modern iteration in Adolphe Alphand's creations for the capital, and thus come to dominate French garden design elsewhere. The Parisian jardin anglais epitomized utility and function, efficiency, flexibility, and adaptation to context. Moreover, its origins in England and its appeal in France at the time of the Revolution invariably connected it to concepts of rights and citizenship, as well as individual experience. The relationship between individual citizens and the newly created greenspaces; the utility of those spaces within the scope of aspirations for the nation, city, and communities; their internal design harmony and sustainability; their association with nature—real or imagined; and their connection to the modern tools of science and engineering technology all figured prominently in the minds and work of the mid-nineteenth-century urban greenspace engineer-developers and helped make Parisian public greenspaces responsive, adaptable, and relevant.

Clearly, parks and squares were not a new urban form at the start of the major renovations of Paris in 1853. What then made them different? Other countries were certainly engaged in efforts to expand public greenspace. British social reformers had pressed for the construction of fully public parks in London early in the century, for example, but those efforts were limited in scale, often relying on patronage and speculation for funding, as in the case of Regent's Park.[17] Planted squares existed in the British capital, as well, but they remained locked and reserved for the wealthy citizens whose property bordered the square. Similarly, there were gardens in Paris long before the mid-century, but with the exception of a brief interlude during the Revolution, those spaces had been restricted and opened only to visitors who were properly dressed or paid subscription fees.[18] Shortly after the city government in Paris embarked on its creation and renovation of urban greenspace, city commissioners in New York City initiated a plan to construct a massive central park in Manhattan. Indeed, in 1858, Alphand shared a detailed report on the construction of the Bois de Boulogne with the New York City officials, which they used in their own planning process.[19] But for all of its benefits, the Central Park was far from the crowded tenements of the city residents who might have profited the most from its open spaces, fresh air, and light. In Paris, however, a concerted effort to assure that greenspace in the form of a municipal park or square existed in every arrondissement of the city including predominantly working-class neighborhoods began during the Second Empire and continued through the early decades of the Third Republic. Therefore, planning initiatives internationally and prior attempts in Paris paled in comparison to the size, scope, and social aspirations of the greenspace development program that accompanied the mid-century redesign of the capital.

The chapters of this book are organized to tell the story of a top-down patriotic and Saint-Simonian-inspired initiative to create greenspaces, which users then took ownership of, pressuring authorities to accommodate their various changing needs and desires. Chapter 1 explores Parisian greenspaces in their national and municipal contexts. It considers the initiators: the planners and the engineers of the park development program. As emperor, Napoleon III had a number of aims and goals for France, some of which he believed might be enhanced through the construction of greenspaces throughout the capital. Public works projects were, for him, both a city project and an expression of a national character. To achieve his goals for Paris, the emperor appointed Baron Haussmann as prefect to oversee the renovations to Paris. Since greenspace in every section of the city was part of the emperor's vision, the task was a formidable one. In 1856 Haussmann created the Service des Promenades et Plantations and called on a young civil engineer with whom he had worked in Bordeaux, Adolphe Alphand, to serve as its director. In their work, the engineers of the park service often aspired to the most accurate representations of what they perceived to be France's national spirit and Paris's municipal character. Although they looked to English garden design for inspiration, they were keen on drawing distinctions between the French and English urban parks and squares.

Some of these ideas realized in greenspaces, therefore, had the potential to reshape Parisians' sense of their city and nation, and France's international image. Parks and squares also helped market the capital city effectively to foreign visitors within an exploding tourist industry as early as the 1850s. Additionally, the city government sought ways to construct greenspace that functioned efficiently within the larger environment of the city and was built with long-term sustainability in mind. This involved establishing support systems to address park construction and maintenance requirements, resources and wildlife management, and human resources development. The city established schools to train landscape architects and engineers, many of whom filled the ranks of the park service. It built nurseries and established tree and turf farms to supply the plants necessary to replenish those lost to damage or blight. The park service set up extensive processes of wildlife management within individual parks and across the system to control deer, waterfowl, and fish populations while at the same time allowing for popular activities such as fishing and feeding ducks, which one city official wrote appeared to have become an absolute "necessity" for park users.[20]

The role of the parks in fostering urban health and hygiene within the city and their contribution to quality of life in the capital is the focus of chapter 2. In the nineteenth-century world of rapid urbanization, planners and residents struggled with issues of pollution, disease, clean air, and *salubrité* as well as the economic viability of modern cities. Reformers and politicians increasingly came to appreciate public greenspace as a necessary and a worthwhile antidote to the unhealthy nature of urban living. Hygienists and doctors particularly interested in the health and welfare of children added tremendous force to this key rationale for greenspace development. Linking parks, children's health and education, and population vigor, these reformers established one of the most enduring and unassailable justifications for the construction of parks and squares in Paris: children's welfare. In so doing, they established a hierarchy of end-users when questions of competing use of space arose, which they did.

The remaining chapters shift attention away from advocates and park builders, rationales and implementation, toward actual use to understand the role that human activity played. Chapter 3, "Greenspace as a Workplace," focuses on those people who worked in the parks and constituted a bridge between the officials of the Service des Promenades et Plantations and the park goers. As a significant presence in the greenspace, park guards, landscape crews, and concessionaires all helped shape and define the spaces through the daily exercise of their jobs. With significant direction from the city, they executed the vision of public greenspace developed by Alphand and the park service. They also inhabited the parks in which they made their livings, as much as and sometimes more than the park goers themselves, and their presence influenced the nature and understanding of the space. They interacted directly with park goers in a way that administrators did not, constituting the public face of the municipal agency. The challenges that park workers faced exposed extant class and gender tensions and the way in which public greenspaces throughout the city could reflect the social geography of Paris. For those who worked in public greenspaces, parks and squares sometimes became complicated social landscapes that were often difficult to navigate.

In chapter 4 the park users as the intended beneficiary of local greenspace become the focus, as well as those whose presence in the space was unwelcome. Key groups of park goers such as the *riverains* (those whose property bordered the park), *habitants* (neighborhood residents, in general), and mothers and children, who had been a primary focus of the park service in its conceptualization of the spaces, actively participated in negotiating and reengineering those greenspaces to suit their own particular needs or desires. Through letters and petitions, neighborhood residents addressed issues of access, design, use, and management. Mothers successfully pressed the park service to reshape the greenspace to meet play and safety needs of their children. Finally, the *proscrits,* or vagabonds, prostitutes, gangs, and criminals, appear in this chapter as the antithesis of the intended beneficiaries. These are not the Parisians for whom authorities planned parks; yet, like other users, they exercised their perceived rights of use of the space, adding to the *popular* nature of the greenspaces as compared to commercial or residential spaces. Chapter 5 similarly explores how greenspaces were used and adapted to meet the challenges of urbanization and a growing mass society. It considers the way in which some groups came together over a shared interest in sports and leisure in the parks. Like those who lived adjacent to or near the parks and squares and others who used them daily, such as mothers and children, they were anxious to lay claim to the space and reshape it to suit their particular use. Skating, cricket, fishing, and cycling were activities that not only worked to foster a much broader kind of community, but also led to organized and concerted efforts to alter greenspaces both on the part of the groups themselves and by the park service in acknowledgment of these emerging interests. These groups ultimately joined with others in a push to protect existing greenspace, presaging twentieth-century preservation movements.

The implications of the alterations to the Parisian cityscape reach far beyond the particular world of nineteenth-century Paris. Our understanding of the place of greenspace within the urban environment is particularly pertinent both historically and at present as contemporary megalopolises such as Istanbul and Shanghai continue to confront some of the same issues planners and city residents in Paris faced in the nineteenth century—issues concerning the concept of public and private space, competing visions of the city, questions about public health, hygiene and salubrité, public participation, environmental viability, and ideas about the nature and quality of urban life. Out of the mélange of challenges and tensions facing planners and city residents in nineteenth-century Paris, a model of a new, modern city emerged, and urban centers throughout Europe and the world, such as Buenos Aires, Montevideo, Alexandria, and Phnom Penh, to name a few, came to mirror the cityscape of

Paris. Along with the wide boulevards and formalized facades, this adaptable yet decidedly French model was easily identified by its abundance of public parks and gardens. These social spaces and cultural constructions tell us much about the society that produced them. As John Dixon Hunt succinctly wrote, "Gardens declare their creators."[21] *Planning the Greenspaces of Nineteenth-Century Paris* argues that nineteenth-century Parisian spaces of verdure had many creators—civil authorities, reformers, planners, engineers, and significantly, city residents and park visitors—and the generative dynamic among those entities defined the greenspace development program and altered the human experience of the urban environment. This study then changes the conversation about mid-century urbanization by demonstrating the critical social, cultural, and environmental role constructed nature played in Paris, and continues to play in a wide variety of urban contexts today.

A NEW GARDEN CAPITAL
NATIONAL PRESTIGE AND MUNICIPAL EFFICIENCY

·••• •••·

[T]hat which he [Napoleon III] understood most clearly of all was that which had to do with executing the great public works projects in Paris, with ameliorating the conditions of the popular classes, with destroying the unhealthy neighborhoods, with making the capital the most beautiful city in the world, all things he desired ardently and about which he never ceased to instruct us.
— DUC DE PERSIGNY, 1868

·••• •••·

At the Congress of Vienna in 1815, the victorious European powers resolved to see that France was sufficiently chastened, and representatives instituted economic, political, and territorial terms under which subsequent French governments would bristle for decades. When Napoleon III came to power in France almost four decades later, he worked to free France from those constraints and the shadow of Vienna. He determined to restore French pride and position as a great power on the world stage, and to make Paris the shining capital of that nation. As part of that effort, Napoleon III aimed to restore internal political order and economic vigor to France. Certainly, a massive rebuilding of the capital could contribute in different ways to both of those goals. He wished to transform Paris and leave his mark on the city. In this, he was following in the tradition of rulers from Philippe Auguste to his uncle, Napoleon I, who likewise altered the spaces of Paris. Like theirs, his aspirations for the capital in many ways reflected his ideological leanings, his sensibilities, and the image of France that he wished to project. He envisioned a grand, prosperous, and orderly Paris. In a speech he delivered in 1850 as president of the republic he told listeners, "Paris is the heart of France. Let us put all our efforts into embellishing this great city, into improving the lot of those who live in it. Let us open new streets, let us clean up populous districts that lack air and light. Let the beneficial sun everywhere penetrate our walls."[1] The municipal government had its own set of concerns about efficiently implementing the emperor's vision. These concerns became part of the transformation of Paris, and they contributed to sculpting an impressive image of the French capital as well as a model of greenspace development.[2]

With this ambitious program in mind, Napoleon III summoned Georges-Eugène Haussmann to Paris in 1853 to take over the position of prefect of the Seine and implement the emperor's vision for a new capital city. Haussmann replaced the sitting prefect, Jean-Jacques Berger. Berger had clashed with Napoleon's minister of the interior, the duc de Persigny, who advocated "productive expenditure" to pay for the redesign of Paris. The city had in its coffers some four million francs, which Persigny recommended using to pay for the interest on large loans that would increase the city's debt, but also pay for the renovation plan.[3] Berger resisted the idea forcefully, saying, "It is not I who will ever borrow the city into ruin."[4] Persigny was undeterred, however, and successfully encouraged the emperor to replace Berger with the prefect from the Gironde, Haussmann. In both Persigny's and the emperor's estimation, Haussmann had proven his loyalty to the regime in 1851 during the uprisings in the provinces following the coup d'état, and two months later when Orléanist supporters protested the 23 January 1852 decree confiscating that family's estates. Haussmann had demonstrated that he was a shrewd administrator by acting quickly and forcefully against the Orléanists and by showing self-control when confronting the much greater threat of the republican opposition.[5] Persigny believed that Haussmann would be the perfect choice for prefect because of this loyalty and because he was unaffiliated with the insiders in the municipal administration, who, like Berger, often opposed the minister. Haussmann was an able and canny administrator and an outsider.[6] There were loose familial connections as well. Haussmann's namesake and godfather was Prince Eugène, son of the empress Josephine and, through his adoption, uncle to Napoleon III, which bolstered the emperor's confidence and desire to elevate the man.[7]

Haussmann was sworn in as prefect at the château of Saint-Cloud, just outside Paris, on 29 January 1853, and he became "the emperor's man" in the capital city. Since the office of mayor of the city of Paris had been eliminated years earlier, Haussmann became the de facto mayor of the capital.[8] He was answerable only to the minister of the interior and the Municipal Council, neither of which gave him much difficulty, since he had the unwavering support of Napoleon III. Indeed, to ensure that none would challenge the prefect's execution of his plan for Paris, the emperor made him a senator in 1857 so that no one in the municipal government could claim to hold a higher position than his.[9] By all accounts, the new prefect was imperious and aloof when dealing with those who resisted the project of rebuilding the capital and its financing, which earned him many enemies. Critics eventually took their revenge when Haussmann's imperial patron became too preoccupied with larger issues to protect him in the twilight years of the Second Empire.[10]

The desire to make Paris a truly grand capital required consideration of what kind of city it, in fact, was and what altering it could or should produce. In the early stages of the public works projects in the 1850s, an article appeared in *Le Monde illustré* concerning the construction of the Square du Temple located in the working-class northeast of the capital. "It is not enough to embellish a city," the newspaper reported, "it is necessary that the embellishing *cleanses*; this is what now ardently preoccupies Parisian city planners."[11] The author discussed briefly the ways in which the changes to the city benefited the health of the young and the old alike, then turned his attention to the way in which constructing the Square du Temple also "cleansed" the dark history of that particular space. "Let us recall the past vicissitudes of this place," he told readers. It was on that spot in the twelfth century, he reminded them, that the Knights Templar, "whose bequest to history is a bloody mystery," had built an imposing fortified castle. During the revolution, he wrote, "the most funereal impressions flew about the place . . . Louis XVI, Marie-Antoinette, Madame Elisabeth, the young prince and his sister spent time beneath its dark vaults, which received many other victims marked for death."[12] The medieval structure was eventually razed in 1811 and replaced by a complex that served many purposes, ultimately as a convent where the sister of the murdered duc d'Enghien spent her final days during the Restoration. Yet now, "vast grassy areas," he concluded, "rocks with tumbling cascades, water features, clumps of trees, and flower beds will smile next spring in these places where time has left such mournful traces."[13] Paris was thus a city with a past that was, in part, dark and divisive. For the author, greenspace would function then as a kind of salve to heal and cleanse the city of these kinds of reminders.

Just as greenery could purge a location of its association with an unpleasant past, a garden could also memorialize a venerable part of the city's history—a history the administration might wish to emphasize. This was, indeed, the case with the Square de la Tour Saint Jacques in the very heart of the city. Germaine Boué, author of a series of publications in the 1860s showcasing the squares of Paris, praised how the city had removed the market stalls that clung to the side of the existing medieval tower on the site and built a new, planted square around its base. The tower was all that remained of the Église Saint-Jacques-la-Boucherie, which had been destroyed during the Revolution. Boué called the project a "rejuvenation" of the historic tower.[14] She applauded the city's painstaking restoration of the statues of saints, which had been vandalized during the destruction of the church. They represented the patron saints of butchers, metalworkers, knife smiths, *aumusse* makers, and paint makers—the professions of those who lived and worked along with other artisans in the surrounding *quartier*.[15] The square also recalled, she noted, the memory of Nicolas Flamel, a wealthy and pious neighborhood resident, who had helped fund the construction of the church and tower, and "lodged the unfortunate homeless in houses built and paid for by him."[16] Quite unlike the small park at the Square du Temple, which obscured a mournful past, the restoration of the Tour Saint Jacques and surrounding square celebrated the talented and hardworking artisans who lived in the center of the capital, and an industrious and generous resident of that quartier. In ways large and small, the increase of greenspace within the city had the potential to affect the city's identity and image.

Alexandre Jouanet, a writer and subsequent attaché to the *Service des Promenades et Plantations,* was an early advocate for the increase in plantings throughout the capital city. He connected greenspace with international prestige. In 1855, he urged the city council to plant more and more trees throughout the capital. "It is in calling to Paris," he wrote, "all of the fruitfulness of large vegetation that we will give to her that character of beauty and of grandeur that is hers in all the world."[17] Five years later he reflected on the work underway throughout Paris in his tract *Paris et ses plantations.* He noted that the great rulers of antiquity went to tremendous lengths to construct, expand, and maintain gardens in their capitals, creating "paradises." Now, he told readers, "[f]ollowing the example of these glorious ages, Paris sees itself accomplishing feats as grand. The great city has its own paradise that we call the Bois de Boulogne."[18] He praised all those responsible for the alterations: "In the hands of enlightened engineers, a distinguished manager, a gardener with great taste, has not the art [of landscape design] there, as in Babylon, displayed all its genius.... And why should it not be so in the capital of a great empire as it was in Babylon and Rome, Sparta and Athens."[19] For Jouanet, the city should certainly be a modern capital, but also a richly planted paradise in the tradition of the cities of antiquity. Whatever the changes or goals, state and municipal authorities needed mechanisms through which to achieve those objectives. The municipal administration required a division that could execute a comprehensive program of greenspace development and make certain that the system would be sustainable. Thus, the Service des Promenades et Plantations, created in 1856 (and praised by Jouanet), and the greenspaces constructed and managed by it, all became the tools of both the nation and the city in a program of national rejuvenation.

Historians have written about Haussmann's ascendancy and tenure as prefect, as well as his fall.[20] Much less has been documented about the formation of the Service des Promenades et Plantations during his time as prefect. In 1854, Haussmann requested that the *Corps des Ponts et Chaussées* (a national corps of civil engineers) place a thirty-seven-year-old civil engineer from Bordeaux, Jean-Charles Adolphe Alphand, at his disposal temporarily to assist with the redesign of the Bois de Boulogne—a request the corps accommodated. It soon became clear to the prefect that the scale of the work in the park and the emperor's larger plan for Paris, which included the increase of greenspace throughout the city, required a new multi-level municipal framework dedicated to public works. Thus, in 1856 Haussmann established the *Service municipal des Travaux Publics,* which was subdivided into three bureaus: *Voie publique* (streets and roads), *Eaux et Égouts* (water and sewers), and *Promenades et Plantations* (parks and squares). Alphand was appointed to head the Service des Promenades et Plantations as its lead engineer and director. In 1867, the Voie publique and Promenades et Plantations merged and Alphand became responsible for both divisions while Eugène Belgrand remained in charge of Eaux et Égouts. On the death of Belgrand in 1878, all three services, and their various subdivisions created in 1871 after the Siege and Commune, came under the sole direction of Alphand until his death in 1891.[21] Thus, the engineer maintained a high-level position in city government long after the emperor and Haussmann had fallen from grace. Indeed, for the last thirteen of his thirty-five years of public service, Alphand controlled *all* aspects of public works projects in the city of Paris, and was in charge of most branches of the municipal administration.[22] Alphand's long tenure as head of the Service des Promenades et Plantations and his broad responsibilities during the Third Republic contributed in a significant way to the continuity that existed in the construction and management of the public greenspaces of Paris for nearly half a century. He had much more direct impact on the development of those urban spaces than did Haussmann, and his vision of "democratized" greenspace—parks and squares that served the largest number of law-abiding citizens at all social levels, equally and well—permeated the ranks of the park service.

Adolphe Alphand was born in Grenoble on 17 October 1817, the son of a colonel in the military. He saw Paris for the first time when he attended the prestigious *Lycée Charlemagne.* Soon afterward he pursued his studies at the *École Polytechnique,* a prestigious, highly selective engineering school, and one of France's distinguished *Grandes Écoles.* Upon graduation from there in 1837, he joined the *Corps des Ponts et Chaussées.*[23] His initial posts were in Isère and Charente Inférieure until he received a commission in Bordeaux in 1840.[24] There he worked on a number of projects related primarily to the harbor infrastructure and the local rail system. While employed in Bordeaux in 1846, he met and married his wife, the daughter of a prominent family, with whom he had three children: two sons and one daughter.[25] His fortune was not great, but Haussmann recalled that "he had an excellent situation in Bordeaux, and he had agreeable relations with the best of society there."[26] The young man's life was so settled and successful that the prefect later wrote of his concern at the time that Alphand might reject his offer of a position in Paris.[27] He did not. In December of 1854, shortly after the birth of his daughter, Alphand moved with his family to the capital and later accepted the new post.

As head of the Service des Promenades et Plantations, Alphand was, in many ways, the antithesis of Haussmann. While Haussmann was arrogant, dismissive, and imperious, Alphand was earthy, earnest, and respected. "When one spoke to him," Georges Lefenestre recalled of Alphand during a posthumous tribute held at the Institute de France in 1899, "he always listened intently and asked questions with clarity although in a slightly mumbling voice; then after several moments of reflection, he responded with but a few words in a gracious tone, yet so firm and clear, despite his accent [*dauphinois*] and so full of conviction and authority, that it hardly left room for any reply, hesitation or resistance."[28] He was a man of simple tastes, "a man of the open air and of sunlight ... who was suffocated and ill-at-ease in the closed, cramped quarters of an office or even a city."[29] Alphand preferred his greenspaces to the *bals* and *fêtes* that Haussmann so famously enjoyed at the Hôtel de Ville. As demolitions began across the city, opening it up to air and light, Alphand had but one thought, Lefenestre recounted, "to conserve as many as he could of these newly created open spaces, purify them with foliage and make them cheerier with flowers."[30] Alphand commanded the respect of others in the government and those who worked under him, while Haussmann seemed to succeed only in making enemies. Yet, Alphand was a loyal friend to the prefect and staunch defender until the end, claiming that Haussmann had been the victim of political intrigues. The government and Paris had shown Haussmann ingratitude, Alphand said publicly in his memorial tribute to his friend at the Académie des Beaux Arts.[31] Of Alphand, Haussmann had written, "we had perfect and constant accord between us, and he was completely devoted to the success of my work. It must be said once and for all what a reliable assistant this eminent engineer was for me, in both character and in talent."[32] This mutual loyalty and admiration had a tremendous impact on the men's professional relationship, and on what Alphand was ultimately able to achieve.

Although great friends and colleagues, Alphand and Haussmann had tastes in landscape design that were as different as their characters. Haussmann's affinity for geometry and the straight line led him to admire the formal *jardin français*. Garden design, he once commented, "attained its apogee when it happily learned how to associate symmetrical, monumental, and sculptural elements, of a character of incontestable grandeur, with those of the palace they appeared to extend."[33] He much preferred the formalism of the seventeenth-century royal garden designer André Le Nôtre to any other style. Alphand, on the other hand, recognized that the free-flowing feel of the *jardin anglais* was better suited to the small public square, making it appear larger and able to serve a variety of users. Instead of André Le Nôtre's vistas and perspectives, Alphand preferred the play of winding paths that moved in and about clumps of trees, one moment hidden from view and then emerging again. These were important, he said, because "it must be counted, among the pleasures of a park, the liveliness that the movement of a group of *promeneurs* creates."[34] Indeed, the emperor shared this preference for the irregular jardin anglais. Haussmann knew well that the emperor and Alphand admired the art of English landscape design, and since he trusted the engineer completely, he gave Alphand a rather free hand in the aesthetic, construction, and management of the greenspaces of Paris, preferring to focus his own attention as prefect on roads, buildings, transportation, and water systems.

As a pragmatic engineer, Alphand recognized the challenges of implementing and maintaining the extensive greenspace development program planned for Paris. He knew that he needed an array of supporting systems. The plan would require, among other things, a tremendous amount of available shrubs and trees. In one of his first acts as director of the Service des Promenades et Plantations, Alphand established a network of nurseries and greenhouses in and around the city. He constructed the first of these in 1855. It was an immense complex in the gardens of the Chateau de la Muette on the rue de l'Empereur (avenue Henri Martin) called the *Fleuriste de la Muette*.[35] Native plants, as well as exotic varieties of broad leaf and flowering plants such as plantains, caladium, canna, and wigandia, which were used in seasonal plantings throughout the park system, were grown there in open beds during the warm season and relocated to state-of-the-art greenhouses during the winter. The original nursery was within walking distance of the main offices of the Service de Promenades et Plantations, and it expanded rapidly over the next few years. By 1865, the nursery compound consisted of the homes and offices of the top management of the park service (including Alphand), stables, barns, a library, a huge orangerie, a potting center, a propagation greenhouse, a grafting greenhouse, seventeen large greenhouses and eighteen smaller ones. The entire area was impressive with its 6,587 square meters of open-air gardens and 6,867 square meters of glass greenhouses at a cost of approximately 400,000 francs.[36] By 1867, the Fleuriste de la Muette was not alone among support operations serving the planting needs of the massive public parks building program in Paris. The city had established several additional nurseries and tree farms in and around Paris. There was a complex near the racetrack at Longchamps dedicated to cultivating deciduous trees; one at Auteuil for coniferous trees, evergreens, and groundcovers; one on the banks of the Marne River where the large plane and chestnut trees that would line the city's boulevards were grown; and one just outside the defensive fortifications at Rueilly dedicated solely to perennial ornamentals.[37]

Among the personnel who worked in the compound were some forty-five students who came to Paris to study horticulture there. Some of the students were French, but the majority came from other countries such as England,

Germany, Belgium, and Holland. Students received small stipends based on merit and remained in training for two to three years before either joining the park service or returning to their own countries.[38] The city expanded its educational facilities with the foundation on 14 March 1867 of the *École municipale et départementale d'horticulture.* Located in the Bois de Vincennes, the school today bears the name of Alphonse du Breuil, professor of agriculture and silviculture at the *Conservatoire impérial des Arts et Métiers* during the Second Empire.[39] These institutions ensured a trained labor pool, helped to create a recognizable standard of public park design and administration, and facilitated the exportation of that model via the many French and international students who studied at them and took up positions throughout Europe. Thus, quite early in the implementation of the Parisian greenspace development program, educational systems emerged that supported the efforts of the Service de Promenades et Plantations and positioned Paris as an international center for the study of landscape design.

While schools and nurseries contributed significantly to creating a sustainable system of park design and construction in Paris, two of the greatest practical challenges for the Service de Promenades et Plantations remained the creation of tree-lined boulevards and the establishment of grassy lawns in public greenspaces. Maintaining large trees in the modern urban environment was problematic because air and soil pollution threatened their success. Even the hardiest trees struggled to thrive in the befouled air of nineteenth-century industrial cities. By the mid-century, however, the London Plane tree, a hybrid of the American Sycamore and the Oriental Plane, demonstrated considerable success in London (hence its common name). The characteristic shedding bark of that species prevented airborne industrial pollutants from penetrating and harming the tree, even in the thick, tarry air of London.[40] Alphand began using the variety with great success in the less polluted atmosphere of Paris. Soil contamination, however, remained a problem. In his 1863 work, *Paris et ses plantations,* Alexandre Jouanet explained how underground gas lines, which ran throughout the capital city, often ruptured or corroded, allowing the gas to seep out into the soil, contaminating it and choking off the tree's ability to obtain nutrients from the soil. This was particularly true in the case of younger trees.[41] While the city worked on developing improved conduits to solve the corrosion problem, it also began to use mature trees regularly in installations across the city. Older trees offered the dual benefit of resistance to gas-related soil pollution and the immediate advantage of plentiful shade with an established look to the design. *Le Monde illustré* praised the practice in its announcement in 1864 that the Square Montholon was nearly completed. "The residents of the *quartier*," the paper reported, "shall not have to wait many long years until the vegetation has acquired enough development to offer its advantages because the trees and plants have been transported [from tree farms] in an adult state. . . . [A] foreigner might easily believe that this public garden, improvised in one month, had been in existence for more than twenty years."[42]

The city's ability to move and install mature trees also helped when diseased or damaged trees had to be replaced. A Service de Promenades et Plantations report in 1875 indicated that, due to a dearth of available large trees in the wake of the Siege and the Commune, when they had been harvested for firewood, the city had to resort to using saplings to replace several trees in the Square du Temple. Those frail trees offered little or no shade to park visitors. The author of the report called for removing four large chestnut trees from the cemetery of Batignolles, four plane trees from the cemetery at Ivry, and three poplars from the cemetery of Saint-Ouen.[43] Such requisitioning of larger trees to provision the square illustrates how adept the park service had become at digging up and relocating mature trees, and the extent to which officials of the agency viewed all greenspace (the service was charged with managing cemeteries as well as parks) as part of an urban network. These transplantations often became public spectacles as crowds gathered to watch specially engineered wagons carry a sixty-meter tree through the streets of Paris and lower it into a carefully prepared hole.[44] So successful was this process that these carts remained in operation well into the twentieth century.[45]

In addition to the challenge of planting and sustaining its urban forest of mature trees, the city struggled to maintain grassy areas in the parks and squares. Lawns were a prominent feature of the jardin anglais style. Yet poor soil conditions and climate in Paris made achieving thick luxuriant lawns extraordinarily difficult. In the Bois de Boulogne, for example, grass grew best on the siliceous sands near the river, streams, and on the well-drained embankments around the lake.[46] In other areas, the soil had to be sifted in what Alphand called a "vulgar" way to remove stones and gravel to make a suitable bed for grasses; even that, he said, "was insufficient to amend the terrain, which was completely deprived of the indispensable elements needed to feed vegetation. The management of lawns was extremely difficult under conditions like these."[47] In addition to poor soil conditions, the warmer, drier climate in Paris as compared to London made maintaining large areas of healthy lawn a significant challenge. Alphand commented on the difference in his work *Les promenades de Paris.* The parks in London, he claimed, were generally expansive, open spaces on which livestock might graze even in the midst of the city. These spaces had fewer trees, and the "open lawns, where a large number of the public could gather, were well-suited to the [cooler, damper] climate of England. They would be less convenient in France."[48] By the time he published his 1873 treatise on landscape architecture, however, Alphand felt the Service des Promenades et Plantations had done a reasonably good job of addressing the problem of grassy areas of the bois. "Today," he wrote, "by having taken care to water sufficiently during the dry season, by spreading a bit of compost, by reseeding bald areas, and by ceaselessly removing weeds, the lawns are maintained in an *acceptable* state."[49]

The difficulty of creating grassy areas in the Bois de Boulogne and elsewhere required expensive solutions. Soil had to be carefully prepared before seeding, then sifted, aerated, and harrowed either by plow or by hand. A precise, somewhat exotic, mixture of rye (imported from England), fescue, tor grass, crested dog's tail, and bromegrass was spread by hand and tamped using heavy cast-iron cylinders that were pulled by plow horses.[50] This extraordinary effort paled compared to the cost. Alphand estimated that the cost of establishing the 273 hectares of lawns in the Bois de Boulogne was 397,800 francs, literally a small fortune at the time. Additionally, the cost of maintaining those lawns, once established, amounted to 880 francs per hectare or 240,240 francs yearly.[51] This effort and expense made the protection of these grassy spaces from damage a high priority for the Service des Promenades et Plantations. Park guards were directed to see that park visitors did not trample and flatten the hard-won grass, or wear paths in it.[52] The city's obsession with maintaining lawns in the public greenspaces resulted in the establishment of turf farms in the 1880s to grow grass for seed and to produce transplantable turf. These farms were located in the broad trenches that were part of the city's defenses by the Porte Maillot. The military permitted the use of the vacant land on the condition that if a defense of the city became necessary, the park service would immediately relinquish to the military any claim of rights to use.[53]

The desire for a more self-contained and sustainable system of supply for greenspace went beyond the flora of the parks and squares to include the fauna as well. There were as many challenges concerning waterfowl, fish, and deer as there were to successful plantings. Sometimes, in the case of wildlife, the problems lay in too many rather than too few. The more sizable parks had considerable populations of waterfowl because the city strictly forbade firearms and poaching in public greenspaces. This caused much consternation for the lead horticulturalist of the park service, Jean-Pierre Barillet-Deschamps. In 1857, when the newly renovated landscape of the Bois de Boulogne was only a few years old, Barillet-Deschamps submitted a report to Alphand requesting the removal of the ducks that filled the large lake of the park. He included a list of the "problem" species, writing, "I have the honor of asking you to kindly issue the necessary orders to clear the large lake of the following aquatic birds—generally all of the species of ducks who wreak havoc and tear up the roots on all the herbaceous plants on the islands."[54] An army of ducks had been plucking away at the very plants the park service had so painstakingly propagated and planted in the spot. There were other birds that ravaged the plant life as well, he added, but they did far less damage than the ducks. "A few hunters," he was confident, "would suffice to break those birds of the habit."[55]

It is unclear what Alphand's immediate response was to this request, but at some point after Barillet-Deschamps's report, the city decided to enter into a contract with one M. Gérard of nearby Grenelle to feed and manage the duck population in the hope that doing so might reduce foraging. That contract and wildlife management strategy resurfaced in 1860 when the park service faced an increasingly serious waterfowl problem in the Bois de Boulogne. The duck population had grown to over 450 birds. The conservator of the park, Auguste Pissot, reported that Gérard had come to see him to say that he (Gérard) no longer wished to hold to the contract that he had with the city for feeding and managing of the ducks on the lake. He wished to know how he might be released from the contract and duly compensated.[56] The agreement had been that Gérard would supply feed for the ducks (at his own expense) and remove sick or dead birds for a set period. At the end of the term, the city would retain the ducks, harvest a percentage for sale at les Halles, and pay Gérard a set indemnity, or Gérard could harvest a percentage on his own to sell in the marketplace. His profit motive had been that he could recover the cost of feed in the revenues from sales, or in the agreed-upon indemnity. Yet by 1860 the size of the population had grown beyond what Gérard or the city had anticipated, and the cost of food had risen so much so that he could no longer profit from either sale or indemnity. He was now seeking an equitable end to this arrangement.

After his meeting with Gérard, Pissot told Alphand that, to answer the request, the park service had first to decide if it intended to clear the park of waterfowl altogether, as Barillet-Deschamps had recommended earlier, or maintain a population there. On this question, Pissot wrote that, for "four years already, these birds have animated the waters of the lake. The public is accustomed to seeing them there. This has become for them [the park visitors], not only an amusement, but I would say that throwing the food to the ducks is almost a necessity."[57] Pissot added that he believed the ducks were a complement to the overall landscape composition. Besides, he doubted that they could be entirely removed. He proposed establishing a new contract with M. Gérard that would suit the needs of both parties. He suggested that the park service should allow Gérard to harvest all but 125 ducks. The species of duck that were allowed to remain in the park would be a mix determined by the park service to maintain diversity. To compensate Gérard for the ducks that he did not harvest, the city would allow him to take forty yearling does from the deer park. This would help address another wildlife population problem since the deer herd in the reserve section had outgrown the size of the range and was increasing yearly. Having thus resolved the former agreement, the city would enter into a new relationship with Gérard wherein the city would pay for his services and the cost of the food. Profits from the regular harvesting and sale of the birds would offset those costs. In this way, the park service could significantly reduce two problem wildlife populations and maintain a small number of birds at the lake for the public's pleasure. Gérard's profit would no longer be tied to the commodity; rather, it would be linked directly the service he provided, and the Service des Promenades et Plantations would take greater ownership over waterfowl management. Such was the learning curve for the agency in similar situations involving wildlife management in the parks. The kind of wildlife that would exist in these greenspaces, in what numbers, and who would manage those populations were all issues that were worked out in the earliest decades of the greenspace development program. Eventually, the Bois de Boulogne became a kind of breeding center supplying newly constructed parks with their own populations of waterfowl to be managed in like fashion.

There were similar issues concerning the fish populations in the various park lakes. Here too, the city used the park resources to generate revenues, which helped, in a small way, to defray management costs. Although the revenue never completely covered the costs of construction and management, it was, for the city, an efficient use of its resources, and managing them thus helped to justify the expense of ostensibly non-revenue-producing use of city property. While the duck and deer populations had exploded quickly in the parks, the number of fish in the lakes and ponds of the Bois de Boulogne and the Bois de Vincennes had always been easy to manage through seasonal netting and sale at les Halles. As in the case of the ducks and deer, authorities considered fish to be a park resource. Indeed, on 2 October 1870, during the Siege of Paris, the government requisitioned the fish in the Bois de Boulogne for the provisioning of the city.[58] In general, fish were much easier than other wildlife to manage through harvest and sale, but, in an effort to satisfy many Parisians' desire to fish in the parks, the Service des Promenades et Plantations instituted a licensing system in 1872.[59] Each year, it granted a limited number of fishing licenses sufficient to meet public demand and keep the fish population in balance. Park fishing increased in popularity toward the end of the nineteenth century, and the number of licenses increased as well as the park service sought to accommodate the public. The parks soon had a problem of too few fish rather than too many, as in the case of the ducks. To deal with the challenge, the park service set aside the Parc Montsouris as a fish hatchery. It prohibited fishing at the lake there and used the fish from Montsouris to restock the other park lakes.[60] In this case, the park system operated as a single mechanism, and internal resources were used to meet internal needs, helping, as with the management of other wildlife, to make the parks as self-sufficient as possible.

Even the physical features in the parks could be utilized to produce moderate revenues for the city. In the winter, sections of the lakes in the Bois de Boulogne and Bois de Vincennes (and later the smaller parks) were regularly cut into for the ice they produced. In the Bois de Boulogne alone, the lakes and ponds amounted to a surface area of fourteen hectares, which could produce a significant amount of ice during a cold winter. In 1857, Alphand ordered the construction of ice warehouses that could store more than ten million kilograms of ice cut from the lakes, keeping it cool well into the summer months.[61] The project was completed in 1859 at a total cost of 408,000 francs. *Le Monde illustré* reported in 1858 that the city's motive for building the icehouses was to use the profit from the sale of ice to make up for the amount it had exceeded the original budget of two million francs allocated for the embellishment of the park. This operation, the newspaper indicated, was not unlike the other concessions established in the parks for similar reasons; yet the sale of ice, it said, would prove to be the most profitable of them all.[62] The operation also helped increase the supply of ice to the capital during the warm season, reducing the price for Parisians.[63]

This kind of internal municipal efficiency squared with Haussmann's view of the modern city—an organism where functionality reigned supreme. It also contributed to enhancing the image of Paris as both a beautiful city and a rational, modern capital worthy of a great nation, something that fit the emperor's aspirations. There were many who shared the sense that the Service des Promenades et Plantations and the greenspaces of Paris had a role to play beyond accommodating park goers. As ardent nationalists, they believed that the public parks and squares offered an excellent opportunity to convey something of the glory and grandeur of France. Just as the gardens at Versailles reflected the *gloire* of Louis XIV, the parks and squares of Paris came to be seen as the embodiment of the gloire of the French nation (be it empire or republic).

In 1865, the well-known sculptor Auguste Ottin championed the idea of turning the planned Parc des Buttes Chaumont into a kind of horticultural museum dedicated to French flora. Charles d'Alleziette, an inspector for the park service, shared Ottin's opinion and wrote to Alphand in support of it. In his report, d'Alleziette argued: "The indigenous flora [of France] is richer in beautiful trees, shrubs, and flowering plants than one generally supposes, and the accidental disposition of the Buttes Chaumont would permit the cultivation there, in various displays, of the vegetation of the plains, that of the mountain foothills, that of the Alpes, and even the vegetation of Provence and of the departments of the Midi."[64] The inspector reminded Alphand that he had suggested the plan to the director a few years earlier in 1862 when he wrote that "[n]ot only would it [a garden planted solely with indigenous varieties] be useful to science and horticulture, it would give satisfaction to national pride in making known all the ignored or misunderstood riches of our France."[65] Alleziette included an extensive list of the plants that might be used in the design. He noted in a postscript that, if the Buttes Chaumont were to be considered too large for such a project, he could easily adapt the design to a smaller park or square.

Alphand, in fact, had already communicated Ottin's idea to Jean-Pierre Barillet-Deschamps, asking him to review the plan and give his opinion on it. Barillet-Deschamps agreed with Ottin that the design and layout of the park did indeed allow for the display of a great variety of plants, but he felt it did not provide appropriate climatic conditions. The suggested plants would neither survive nor be aesthetically pleasing. "Vegetation of French origin [that could be used in such a way] almost all have small leaves and are of a nearly identical structure," he wrote, indicating that such a display would appear monotonous.[66] He also pointed out that French oaks had been tried already on the quai des Tuileries, and even the most hardy had not survived "the dust and gaseous emanations of the city."[67] As a horticulturalist, Barillet-Deschamps was more interested in using greenspaces to educate the public not about the riches of France but about the great wonder and diversity of plants in general. He recommended smaller displays of attractive plants from places such as Japan, Australia, or the Americas. Neither Ottin's nor Barillet-Deschamps's plans ever came to fruition, and the park service continued to balance form and function, planting the parks and squares with varieties that were sturdy enough to be cost effective, and attractive enough to offer a pleasing appearance.

Greenspace development in Paris had captured the attention and interest of those outside of the capital as well. Its highly visible early successes in the creation of a multitude of new open spaces and the way in which the Service des Promenades et Plantations functioned efficiently within the larger municipal system made Paris a model for park development throughout France and the world. In a lecture delivered at the Hôtel de Ville in Troyes in 1866, Alphand's assistant, Édouard André, trumpeted the successes of the various park constructions in Paris and told of projects in the capital that were still in the planning stages. He spoke of the way in which Paris had become an inspiration to other cities throughout France. "At the instigation of the city of Paris," André told the audience, "the great cities of the provinces wanted to have their own public gardens and break free of the routine in which they have slept for so long."[68] As examples, he cited the new park in Lyons designed by Eugène Bühler, which he described as Lyons's own Bois de Boulogne. In Marseilles as well, he noted, the municipal government had built an extensive park despite the challenges a seaside climate posed to its vegetation. "Rouen, Montpellier, Avignon, Lille, Tours, Angers, Caen, Nantes, Strasbourg, and Troyes," he told them, "have been endowed with public gardens for which the municipal governments have made bold sacrifices, repaid by the warm welcome of the public."[69] It was, according to André, nothing short of an urban-planning revolution and the rejuvenation of France. "This regeneration is now in full force. It is to be believed that it shall not stop there, to the greater good of the populations who are the object of it."[70] Indeed, cities throughout the country were anxious to follow the lead of Paris by creating public greenspaces. Earlier that same year, *Le Monde illustré* reported the recent dramatic alterations to the city of Rouen: "In the emulation that now leads the cities of France, large and small, to cleanse and embellish their streets and their neighborhoods (Paris in this genre, as in everything, is the model), Rouen, the old Norman capital has not been left behind, far from it."[71] Among the many changes, the author described a new boulevard and the creation of a large, planted square in the midst of the city. "The street is called the rue de l'Impératrice; the square, the jardin Solferino—one a charming name the other glorious" referring respectively to the Empress Eugénie and France's military victory in 1859.[72] This kind of effusive national and civic pride permeated these discussions of greenspace development inspired by Paris.

Cities in the provinces continued to model themselves after Paris, at least in the configuration of urban spaces and in the construction of public greenspace. The engineering and natural science journal *La Nature* reported in 1874 that the construction of greenspace in the city of Paris, which continued in its effort to place parks and squares in every quartier, had been emulated throughout the countryside. "If one glances over a plan of Paris," the paper wrote, "one notices that the squares are spread out with a well-determined goal in mind: each arrondissement, except two or three, possesses an important public garden.... The example that Paris has given has been followed closely by cities throughout France: Lille, Marseilles, Nice, Lyons, Bordeaux, etc."[73] In Bordeaux, in fact, the influence of Paris as a model resulted in conflict among municipal council members in 1887 as different interpretations of the model vied for dominance. In 1880, the municipal council voted to purchase the private gardens of Longchamps, known as the Parc Bordelais, which at the time were only open to subscribing members or for an entrance fee. The purchase price of the property was 350,000 francs, and the city leaders intended to turn the garden into a fully public park (without any entrance fee). After two years of haggling over how to pay for the property, the council finally agreed to borrow the necessary funds, and the measure passed in 1882 by a vote of eighteen to two with two abstentions.[74] The goal of those supporting that purchase, as articulated by municipal counselor Merillon during the session, was to establish a public park on the site so that the "working-class population could find there on Sundays a place of relaxation that today they are obliged to travel far and along dusty roads to find."[75]

Five years later, the Bordeaux municipal council met again to vote on a project to overhaul the fence and gates of the park.[76] The height and configuration of the fences and gates were patterned after those of the Parc des Buttes Chaumont in Paris, although of a simpler design with the exception of the main entrance, which was to be substantially grander and more expensive. Those in favor of the measure argued that the importance of the entrance required the more expensive design, and besides it would last for quite some time, and future administrators would one day commend them for their foresight.[77] Those opposed, led by municipal counselor M. Olaqnier, were in the minority. They argued that the other members had forgotten the original purpose for purchasing the property. The city obtained the garden originally, Olaqnier claimed, "for the creation of a *promenade populaire*. In all the votes that the council had taken on the matter, this democratic idea has dominated—the idea of creating an open park, accessible to even the most humble."[78]

Olaqnier reminded the council that the description of the original design for the park submitted by the landscape architect, Eugène Bühler, was modest and lacked all pretension. "Many members of the council," he told them, "remain faithful to these ideas and would like to give this new project [the gate] a character eminently popular and democratic." Olaqnier saw the proposal of an ornately decorated gate as the first move to curtail the right of all to enter the park freely. He continued: "If you give this entrance such an ostentatious character, are you not afraid of putting off the timid, the *gens du peuple*, the workers, who would be coming fresh from the workshop to breath the pure air of the country? If you give such brilliance to the exterior decoration, might you not determine to forbid entrance to the park to those impoverished persons you find too modestly dressed. . . . [I]f you come to make the Parc Bordelais the rendezvous of the elegant faction of the population, you take this countryside away from the workers who have none, something you have promised them for many years."[79] Olaqnier asked the council to send the design back to be brought into line with a more egalitarian conception. For him, the model Paris presented to the provinces of park development was primarily one of providing access to greenspace to all citizens, rich and poor, while his opponents on the council saw in that same model primarily an aesthetic expression of civic pride. In the end, Olaqnier lost his bid to keep the park entrance as unpretentious as possible, in a vote of twelve to ten with four abstentions. The debate in Bordeaux illustrates that the fullness and ethos of the Parisian model was not always translated to the provinces easily or intact, particularly given the nature of local politics and governance. Provincial cities generally lacked an administrator with the kind of personal vision, experience, strength of reputation, and actual power over city public works projects that Alphand had in Paris.

Along with ebullient admirers, there were also harsh critics of the Service des Promenades et Plantations and of the changes to the cityscape of Paris. Journalist Victor Fournel, a tireless opponent of Haussmann and the empire, took aim at Haussmann and the work that the park service had completed in the capital. In an article in *Le Correspondant* in 1864, he lambasted the prefect for having "rudely attacked" the "pride and joy of Paris" by reducing the size of the Luxembourg Gardens in order to build the rue de Médicis.[80] "Neither the petitions of the residents of the *quartier,* nor the unanimous vote of the Senate could save the Luxembourg from this affront," he claimed.[81] Moreover, he said, there were rumblings concerning the fate of the *Pépinière,* a medieval nursery complex on the southeastern end of the gardens.[82] "The shadow of Haussmann," Fournel told readers, "obsessively pursues the peaceful visitors to the garden. Around every corner, they believe they glimpse the menacing silhouette of an engineer armed with his plans and taking his measurements. Until they [city administrators] complete the embellishments to Paris, the sword of Damocles hangs over their [park visitors'] heads."[83] Fournel continued his scathing criticism of many of greenspace development projects: "Except for the earth and trees ... all is artificial in the Bois de Boulogne."[84] As for the Bois de Vincennes, *it* escaped similar embellishment perhaps, he believed, because it was in a plebian neighborhood. Worst of all, according to Fournel, was what the park service had done to the Parc Monceau. The city had reduced "the most beautiful garden in France" to the level of a "vulgar square."[85] It was boring and cold now, and perhaps the most unforgivable assault, as far as Fournel was concerned, on this jewel of the eighteenth-century aristocracy was that the city had divided and sold as lots some of the land on the periphery of the original park. He concluded with a direct attack on Alphand. "This is what the official newspapers call the *beautiful creation* of M. Alphand. I say that this pretended *creation* is an *execution*."[86]

Despite the trenchant criticisms and all that might be lost of the original model when provincial municipalities tried their hand at fashioning their cities after Paris, the example the capital set for greenspace development and the image it portrayed of the nation remained strong and persuasive. The expositions, which attracted thousands to the capital city, became the best way of promoting that model, and by extension France. They merged national image and municipal efficiency on a highly visible international stage. Britain's exposition of 1851, which featured the Crystal Palace, had set the bar high and transformed the trade fair into the grand Universal Exposition, a tool of national promotion and international competition. Although Paris had hosted smaller expositions, the *Exposition Universelle* of 1867 was by far the most ambitious expression of the event in France at the time. It was Paris's moment to shine, and France's opportunity to burnish its image among the nations of Europe. Greenspace played a key role in showcasing the city, as well as the nation, during the months of the exposition. Indeed, the inauguration of the new Parc des Buttes Chaumont was carefully scheduled to coincide with the opening of the exposition at the Champ-de-Mars on the first of April. Workers rushed to complete the task of turning the dismal site of former quarries and slaughterhouses at the buttes into a fine example of the jardin anglais style in time for the exposition. Unfortunately, on opening day, neither the park nor the exposition grounds were completed.[87] *Le Monde illustré* reported on 4 May 1867 that "[t]he Parc des Buttes Chaumont was almost finished."[88] Visitors had access to certain sections of the park, the newspaper told its readers, and "in eight days hence, the Parc des Buttes Chaumont will be able to show its splendors to foreigners, drawn to Paris by the great Exposition of 1867."[89] Europeans had heard much about the renovations to Paris, and the municipal (and national) government was anxious to demonstrate France's new engineering prowess as applied to public greenspace. The city parks were on display as much as the wares and artisanship in the exposition hall. This positioning of Paris as *the* image of France persisted even when the expositions were held elsewhere. In 1874, when London hosted the International Exposition, the French contingent constructed several models of the city of Paris showing the various improvements to it as a result of the public works projects of the last two decades. *La Nature* raved about the display and told its readers, "Today, in order to truly understand what is modern Paris, the best thing to do is go to London."[90]

When held in Paris, the expositions became part of a concerted effort to promote not only the nation but tourism as well. Millions of exhibitors and visitors flocked to the city. In a fundamental way, tourism paid for some of the cost of the changes the prefect had made, and it became his oft-used defense of those renovations. Reflecting on the period, the conservative commentator Maxime du Camp wrote in 1875 concerning Haussmann's argument: "He wasn't wrong—the more Paris was made vast, airy, and magnificent, the more foreigners would be drawn there, would visit and bring their money which became be a source of prosperity to the population. Certain constructions, that initially seemed to be but luxurious follies, rendered five times the amount they cost because they attract foreigners who stay amongst us."[91] The city and the parks were not only a reflection of France; they made a significant contribution to economic prosperity. As part of the dynamic of improvement, attraction, promotion, and profit, the spaces of the capital became crucial to France's status in the international community and its desire to participate in an increasingly lucrative tourism industry. This was certainly the case in 1867, when several of the most eminent writers and personalities in France participated in producing the *Paris Guide,* a massive two-volume compendium of essays published in concert with the Exposition Universelle. Essays focused on various aspects of science, art, and daily life in the capital city. Contributors to the project included politicians and writers such as Léon Say, Victorien Sardou, Pierre Véron, Alfred Delvau, Arsène Houssaye, Jules Favre, Georges Sand, Édouard André, Eugène Pelletan, Maxime du Camp, and Victor Hugo, among others. It was an effort to describe Paris, Parisians, and by extension, France to the flood of visitors attending the exposition—a reintroduction, as it were, of a great and glorious France to the world. Given its release date, it stands as both a crescendo to the *fête impériale* of the Second Empire and the beginning measures of an overture to the Third Republic. It encapsulated the interconnectedness of contemporaneous ideas of national image and national pride, the capital, urban spaces, and rejuvenation.

In his essay in the *Paris Guide,* Eugène Pelletan boldly and chauvinistically expressed what many Parisians, Frenchmen, and even the emperor in 1867 would certainly have eagerly endorsed. In so doing, he highlighted the extent to which capital cities, in the second half of the nineteenth century, came

YBP Library Services

HOPKINS, RICHARD S., 1961-

PLANNING THE GREENSPACES OF NINETEENTH-CENTURY
PARIS.
                                  Cloth   218 P.
BATON ROUGE: LOUISIANA STATE UNIV PR, 2015

AUTH: WIDENER UNIVERSITY. HISTORY OF PUBLIC
PARKS, SQUARES & GARDENS IN 19TH CENT. PARIS.
LCCN 2014035918
  ISBN 0807159840   **Library PO#** GENERAL APPROVAL
                                    List   42.50 USD
  5461 UNIV OF TEXAS/SAN ANTONIO   **Disc**   17.0%
  App. Date 8/19/15 URB.APR   6108-11 **Net**   35.28 USD

SUBJ: URBAN PARKS--FRANCE--PARIS--HIST.--19TH
CENT.

CLASS SB484   DEWEY# 712.50944361 LEVEL ADV-AC

YBP Library Services

HOPKINS, RICHARD S., 1961-

PLANNING THE GREENSPACES OF NINETEENTH-CENTURY
PARIS.
                                  Cloth   218 P.
BATON ROUGE: LOUISIANA STATE UNIV PR, 2015

AUTH: WIDENER UNIVERSITY. HISTORY OF PUBLIC
PARKS, SQUARES & GARDENS IN 19TH CENT. PARIS.
  LCCN 2014035918
  ISBN 0807159840   **Library PO#** GENERAL APPROVAL
                                      List   42.50 USD
  5461 UNIV OF TEXAS/SAN ANTONIO   **Disc**   17.0%
  App. Date 8/19/15 URB.APR   6108-11 **Net**   35.28 USD

SUBJ: URBAN PARKS--FRANCE--PARIS--HIST.--19TH
CENT.

CLASS SB484   DEWEY# 712.50944361 LEVEL ADV-AC

to communicate something of national characters. Disarmingly, Pelletan first insisted that it was, of course, "childish" to think about which capital or nation is the greatest when each has its own particular strengths. Still, he asked, "when one takes a look around Europe and one looks for the city that represents the general best, it is not London, which is but a marketplace; it is not Berlin, which is but a university; it is not Vienna, which is but a concert; it is not Florence, which is but a museum; and it is not St. Petersburg, which is but a garrison. Who is it then if it is not the city that is at once commercial, industrial, poetic, artistic, literary, wise—in a word, the city of Paris. . . . [I]t is true then that if each people had to name a capital of Europe, they would place their finger on Paris and they would say: La voilà!"[92]

Although a staunch critic of Napoleon III's regime, he was, as were others, transported by nationalist sentiment and civic pride in anticipation of the throngs who would visit Paris and the Exposition Universelle that year, and in years to follow. This notion of Paris as the greatest city in Europe and capital of the civilized world helps to explain how the experiential and physical spaces of the capital came to be understood as tangible expressions of national character and linked to national prestige. For some, great capitals had to be carefully conceived and constructed to reflect a particular vision of the city and nation. To be a truly great capital, Napoleon III believed that Paris had to be orderly, modern, economically prosperous, and fulfill his understanding of the revolutionary promise of equality. For Jouanet and others, the capitals of antiquity were recognizable because of their lushness, their planted paradises; thus Paris needed greenery to assume its rightful place among them. The mid-century greenspace development program then easily fit these varied conceptualizations of the capital and could facilitate some of the goals of the state. The Service des Promenades et Plantations, dominated and run by engineers interested in functionality (sometimes nearly obsessed by it), demonstrated a remarkable ability to set up sustainable systems of internal operational efficiency in order to reshape the cityscape. This particular quality, combined with a clear leadership vision and stable management over time contributed to the success of translating the greatness of the nation into an urban planning and greenspace expansion program that made Paris a model, placing it at the international forefront of landscape design. Prestige and efficiency, however, were not the only motors driving the renovation of the city and the construction of parks and squares, or helping to determine success. In the context of nineteenth-century France, the idea of the modern city became inextricably linked to healthy living conditions for Paris's exploding population. Napoleon III's great imperial capital city had to be a healthful and salubrious city as well as grand and expressive. The parks and squares so efficiently built, renovated, and managed under the direction of Adolphe Alphand would become central to fulfilling that part of the vision, as well.

PUBLIC HEALTH AND THE GREENING OF PARIS

•••• ••••

The tree is at once an instrument of aspiration, of filtering and of disinfection. From all of these points of view, uncontestable benefit.
—JEAN-BAPTISTE FONSSAGRIVES, 1874

•••• ••••

Napoleon III was just as interested in the ideas of the utopian socialists as he was in creating a grand capital that would reflect France's national character and aspirations. His essay "The Extinction of Pauperism," written while in exile, stands as a remarkable (although some would say extremely flawed) expression of Saint-Simonian economics. For many Saint-Simonians, public health and urban renewal figured prominently in much of their philosophy. Unlike other utopian socialists such as Charles Fourier and Louis Blanc, who championed egalitarian, autonomous workers communities (non-industrial in the case of Fourier), Saint-Simonians envisioned a technocratic state based on expertise within which the industrial city in general, and Paris in particular, functioned as a center of trade and human exchange. There science, technology, and especially public works could be employed successfully to assure healthfulness, civic order, and cooperation among citizens.[1] This concept of health and prosperity linked to a reconfiguration of the built environment of Paris constituted a significant part of the emperor's goals for the capital city. At the same time, new scientific research in chemistry, biology, and sociology complemented a new philosophical approach to using science to ameliorate the conditions of modern living—positivism. The positivist appreciation of the role that scientific study might play in addressing social and political challenges fit well with utopian-socialist concepts of society. In the context of what was known about the spread of disease in the mid-nineteenth century, what was believed true about the function of green plants in the urban milieu, and in light of the positivism that redefined the social questions of the day as those the experts, scientists, doctors, politicians, and engineers could best address, the parks and squares that figured so prominently in the redesign of Paris were a key element of a public health and welfare initiative. Begun in this way, nineteenth-century comprehensive greenspace planning and development took on a decidedly positivist hue, which continued as the program was expanded during the years of the Third Republic.

The location of these spaces throughout the city, along with a rationale rooted firmly in the notion of an intrinsic right to a healthful environment, contributed to an increase in the "publicness" of municipal greenspace. Yet the path was by no means a direct one. Questions of the power and responsibility of the state to address social issues and uphold citizens' right to air and light without subverting the ownership rights of private property muddied the waters, complicating reforms but increasing the attractiveness of addressing health issues through remaking the public spaces of the city. Emerging out of economic hardship and concerns over cholera and insalubrious housing, and linked to science and reform through positivism, public greenspace development came to be associated with necessity, rights, and justice in the many tracts and treatises of the 1850s and 1860s. This remained the central rationale and driving force in the construction of public greenspace through the end of the century. Since the definition of what constituted "health and welfare" evolved over time, so too did the argument for parks construction. As the health and well-being of children became the overwhelming concern of an increasingly pronatalist society, greenspace was linked to national concerns in a new way.

In 1832, the deadly cholera epidemic that swept across Europe was particularly devastating in cities like Paris where the urban population had been steadily increasing and where housing, notably among the city's poor and working classes was cramped and unsanitary. Nearly 18,000 Parisian residents died in only a few months of the epidemic. Although doctors, administrators, and the press expressed a great deal of consternation, they managed to accomplish little in the way of effective change, partly because of prevailing attitudes toward the populations that had been hardest hit, the indigent and working poor, and partly owing to professional differences and a lack of cooperation.[2] In 1849, cholera returned to the capital and the death toll was similarly high. In the absence of any clear understanding of what exactly had caused cholera or how the disease was spread, scientists, doctors, journalists, and reformers now came together with renewed intent and tried to understand the disease in the context of their own expertise and interests, and in terms of what coordinated efforts might be effective against future epidemics. Some, like the Saint-Simonian utopian socialists, focused early on an ambitious approach, one that would alter the public spaces of the city. In 1832, Stéphane Flachat, an engineer and a leader in the Saint-Simonian movement, wrote an essay titled "Le Choléra— Assainissement de Paris" which was published in a collection of essays by various proponents of the philosophy outlining many of the key precepts of Saint-Simonianism. In his essay, Flachat articulated a specific set of recommendations for the city, which would prevent future outbreaks of contagion. He asserted that the municipal authorities ought to complete the *percement* (opening up) of the rue du Louvre (rue de Rivoli) to the Bastille, a plan developed by Napoleon I but set aside because of the Russian campaign in 1812.[3]

Additionally, Flachat called for improvements to the sewer system and flushing the streets on a regular basis. He suggested the prohibition of mooring along the quayside; essentially, moving the port of Paris to an area outside of the city. This would encourage the workers whose jobs had been attached to the shipping industry to move outside of the city as well. There, he believed, they would find for the same price lodging that was more healthful and better ventilated.[4] This idea of the benefits of fresh and clean air, *salubrité*, appeared often in reformist writings.[5] In a government commission report conducted the same year, a committee of physicians, scientists, and administrators came to a similar conclusion. They suggested looking across the channel toward neighbor, and sometimes rival, industrial Great Britain for inspiration. The report concluded that the city administration ought to "borrow" from the British the practice of building urban, planted squares, but place them in the most crowded neighborhoods within the city, where they would offer "an *airy* and healthful refuge for the elderly and for children."[6] In the committee's proposal, as in Flachat's, the means of improving the salubrité of the city had much to do with exterior common spaces.

Along with Flachat, reformers such as Louis-René Villermé and Alexandre-

Jean-Baptiste Parent-Duchâtelet criticized decrepit environments and pointed out the relationship between poor living conditions and public health issues.[7] They blamed the environment rather than the impoverished inhabitants who had few options but to live in such squalid neighborhoods, unlike those who blamed the poor themselves. Housing quickly took center stage for these reformers, and they circulated a series of treatises, studies, and *enquêtes* in the years following the 1832 cholera epidemic and after its 1849 reoccurrence.[8] Not all commentators who were offering solutions in the wake of the cholera epidemics were interested purely in disease prevention. Some, such as Sir Edwin Chadwick and Henri-Antoine Frégier, advocated utilitarianism, moral reform, and crime suppression.[9] Even they, however, acknowledged that the environment either contributed to or exacerbated the issues about which they were most concerned. Improve the physical environment in which the poor lived, they argued, and there would be a decrease in the incidence of disease and likely a decrease in vice and crime. Although these kinds of arguments and discussions were by no means new or unique to the 1830s and 1840s, the incidence of cholera increased the urgency, focus, and impact of these works.

One such study appeared in 1838. Henri-Antoine Frégier, then *chef de bureau* of the Prefecture of the Seine received a prize from the Institute of France for his study of the urban poor entitled *Des classes dangereuses de la population dans les grandes villes et des moyens de les rendre meilleures,* which was published for a wider circulation in 1840. As a police administrator, Frégier was primarily concerned with the reduction of crime within the city, which he believed originated in the impoverished neighborhoods; thus his portrayal of the poor was disparaging indeed. Yet his treatise also attempted to address the problem of inadequate healthy housing that plagued Paris in the decades following the cholera epidemics. He perceived moral degradation to be at the root of criminal activity, and poverty with its deplorable, unhealthy living conditions as the cause of that moral decline. He condemned the housing situation of the poor, blaming it, not them, for the vice that he believed pervaded those populations, and for the "insalubrité publique" that was another result of poor housing. The "comfortable" classes, according to Frégier, found in their dwellings a haven that conformed to their needs, an "agreeable retreat." The poor, on the other hand, driven by economic necessity, nearly always found themselves housed in poorly built, spatially insufficient, and unhealthy lodgings. It was what he called "the harsh law of necessity."[10] If this dynamic could not be altered, and

Frégier believed it could not, it might be minimized by constructing decent housing for the poor. Such clean, salubrious housing would offer the dual benefits of improving public health and, as was Frégier's concern, promoting "bonnes moeurs" among the poor, thereby reducing crime.[11]

In making his case, Frégier cataloged the many professions of the poor and described their living conditions. Of the seasonal workers who flocked to Paris from the provinces each year to work on construction sites and who housed themselves in cheap crowded rooms to save on their expenses, he wrote, "Isn't it regrettable that these brave men sleep so crammed together in little cubbyholes. Accustomed to working in the open air, the cramped nature of their lodgings must be more painful to them than to all others."[12] The *chiffoniers* (ragpickers), according to Frégier, lived in even more "distasteful" conditions. They were likewise crowded into small, often windowless rooms, without regard to gender, and surrounded by vermin, living and dead.[13] A dearth of salubrious housing, he argued, thus threatened not only public health but also civil and moral order. It had become a question of rights, he asserted, writing, "[I]t is impossible to not think upon the necessity to bring about an efficacious remedy to a situation so contrary to the rights of humanity and to civilization."[14]

Michel Lévy, a doctor and professor of hygiene, echoed some of Frégier's concerns just three years later. However, as a physician, Lévy focused more on the spread of contagion than on the root causes of crime. In his *Traité d'hygiène publique et privée* published in 1844, Lévy wrote in a chapter dedicated to the question of habitation, health, and fresh air that, if one were tempted to call large cities in France the tombs of mankind, it would be after they had seen those neighborhoods where "poverty stagnates in houses deprived of air and light, poisoned by mephitic refuse, made hideous by dilapidation and vermin; our most flourishing cities have their cesspools, less affordable than the tent of the Arab, more filthy than the smoke-filled hut of the Polynesian."[15] Steeped in nineteenth-century chauvinism and racism, Lévy's description nonetheless articulated some of the same issues that troubled Frégier: the problem of air circulation in Parisian housing, and the economic factors that exacerbated the situation. He recognized that in multi-unit dwellings there existed almost always a "parsimony of space." These urban structures privileged the exploitation of space over the importance of salubrité.[16] High rents exacerbated the problem as the "cupidity" of owners further reduced shared spaces such as stairwells and landings in an effort to create more revenue-producing space. These new units were most often "deprived of light, damp, and poorly ventilated."[17]

Unlike Frégier, whose general call for better housing lacked specifics, Lévy, in spite of having no training in engineering or architecture, ventured to make concrete and detailed suggestions about how to design and build proper, healthy domiciles. He detailed precisely the kinds of construction practices that would improve the amount and quality of air within buildings, which would in turn improve living conditions for urban residents and reduce the risk of the spread of disease. His comprehensive plan addressed private *and* public spaces, the placement and orientation of buildings on a site; construction materials, layout, wall treatments, heating and lighting; the relationship of the building to the street and street design; arrangement of support structures; and the design of toilets, all with the expressed goal of increasing the circulation of air. Refusing to limit himself to the interior private spaces of the home, he included recommendations for the disposition of alleys, courtyards, and the city streets. He asserted that streets were "canals of air into which pours human stench from every opening of the buildings on either side." Thus, constructions ought to be set back from this "miasmatic flow." Enlarged and open courtyards would encourage foul smells and dangerous emanations to flow upward and outward, improving the air quality of shared living and public spaces.[18]

Although Frégier's interest and ideas about the origins of crime limited his study to the housing of the poor, Lévy argued that the problem of a lack of sufficient, fresh air in cities and urban dwellings was an issue that crossed class lines and threatened wealthier Parisians in equal measure. Richly furnished apartments, overfull with objects of luxury, he claimed, similarly decreased the amount of necessary interior fresh air. "[T]he middle classes," he wrote, "deprive themselves either through avarice or through negligence of necessary space in their domestic installments: nothing is lacking in the gilt boudoirs, in the richly draped alcoves, in the sumptuous studies, except the air that supports life. They know how to use wealth to obtain all types of comforts; but they forget the element essential to health, that is quite simply a sufficient amount of ambient air."[19] This might have been exaggeration on Lévy's part; yet he was, quite cannily and pointedly, communicating that insalubrious housing was not exclusively the problem of the poor. In so doing, he attempted to personalize the problem for those in better economic circumstances, broaden his call for building and planning reforms, and, unwittingly perhaps, help pave the way for further discussion of the essential and universal human need for fresh air in the city environment. For Lévy, a healthy abode with a generous and free flow of fresh clean air was not merely a right; it was in fact a physiological necessity.[20] The air around us, he concluded, was "the field of life." But housing construction, he wrote, continued to "answer more the needs of civilization than to the exigencies of nature; it serves but to shelter man periodically and temporarily; if he stays too long in a fixed abode, he impairs the conditions of his organic life."[21] Linking air and health is ubiquitous in the published reports and essays following the cholera epidemics; in this way the two became virtually synonymous.

Lévy's work reflects the influence that positivist philosophy had on many educated reformers and planners in the decades leading up to the 1850s and throughout the Second Empire. Initially associated with justification for authoritarian rule, positivism later became linked to republican opposition, owing to the number of positivist scientists and intellectuals who were part of the ongoing opposition to the regime.[22] Lévy along with other social reformers mixed actual scientific experimentation with social-science observation and hypothesis. His work on public and private hygiene draws on much of the scientific research that was contemporaneously underway concerning air composition, the physiology of breathing, and disease to make his case for reform. Before one could know how to build housing and plan streetscapes, he argued, one needed to know precisely how much fresh air was necessary to live, promote health, and stave off disease. Science could provide these kinds of answers. Lévy cited the experiments of many of his contemporaries such as Michel Chevreul, Jean-Baptiste Dumas, Jean-Baptiste Boussingault, Jean-Claude Eugène Péclet, and Pierre Adolphe Piorry as well as notable earlier work, such as that of Nicolas LeBlanc, enthusiastically detailing and utilizing their findings to bolster his claims about the need for the circulation of clean air in the urban environment, and the relationship between fresh air and disease prevention.

The noted pathologist Antoine Ambroise Tardieu shared Lévy's conviction that science held the key to solving many of society's ills. In his *Dictionnaire d'hygiène publique et de salubrité* under the entry concerning *assainissement* he wrote, "In the interior of dwellings, the commensurability of capacity with the number of inhabitants, the exchange and refreshment of confined air, the absorption of foreign elements that can become mixed in the air supply, constitute the first conditions of sanitation. Science possesses the means to fulfill those conditions."²³ Tardieu's faith in science was unmistakable, but it was not a blind faith. He understood the limitations. "As to the causes of cholera," he wrote, "it is important to recall that pestilent diseases are not those that man may be given to penetrate the origin of or to know the principle.... There, in the words of M. Littré," he continued, "all is invisible, mysterious; everything is produced by forces of which the effects alone are revealed to us."²⁴ Tardieu's reference to Émile Littré is noteworthy since Littré was a follower of Auguste Comte and a leading proponent of positivism at that time. This kind of melding of chemistry, physiology, philosophy, and social science—together a kind of science tinged with a moral imperative—as applied to urban planning was part of a Saint-Simonian legacy that, in the wake of the cholera plagues of 1832 and 1849, became a prominent aspect of the justification of Napoleon's park building program throughout the decades of the Second Empire.

Throughout these kinds of published opinions as to what could or might be done regarding the insalubrious housing of the poor, the tension between landlords' private property rights and the state's public health concerns emerge as an ongoing, nagging problem presenting an obstacle to amelioration. The tone varied from benign resignation about the nature of market forces to a real sense of indignation and contempt directed at owners of rental property. Frégier bemoaned the fact that legislation concerning rental properties often tended to focus on the right of a property owner to let out rooms rather than the actual condition of those rooms. He believed that the 1832 ordinance, which extended article 475 of the Penal Code concerning the registration of numbers of guests at inns to include those who became landlords either "habitually or accidentally," was a good thing from the perspective of "public security and tranquility," but it did nothing to address the health conditions in the interior of these rental spaces. He speculated that a general desire to refrain from doing anything that might infringe on the landlord's rights of ownership of private property prompted insufficient measures in the law and the ordinance. This "left without remedy a state of affairs very injurious to the health of lodgers, and could, in the case of a reappearance of cholera, markedly increase its deadly reach."²⁵ It would require a wise administration to foresee the problem and act on it. "The task is difficult, without a doubt," he wrote, "but why not approach it with courage; why permit such *foyers d'infection* to exist in Paris, without making any effort to destroy them?"²⁶ In 1854, Tardieu similarly complained that, although science and the law had come together in recent times to monitor unhealthy housing and industry, and although this had improved the essential conditions of human life in Paris, "private interests" frequently placed obstacles before some of the most useful public health measures.[27] The tension between the rights and responsibilities of individuals and of the state with regard to public health and welfare continued and would be debated for decades.

The question of who bore the responsibility for improving living conditions remained unclear in France and in Britain, where similar debates followed cholera's devastation there. In 1850, British architect Henry Roberts published a book titled *The Dwellings of the Labouring Classes* and submitted it to the Royal Institute of British Architects for consideration. That summer, Louis-Napoleon Bonaparte, as president of the republic, ordered a translation of Roberts's book into French. In a mass circulation, the government sent out copies to hundreds of mayors and city councils throughout France. The following year a copy was also presented to the Bibliothèque Nationale de France.[28] Roberts was not interested in the relationship between housing and disease, or even in specific questions of urban planning. Rather, he was interested in how to design healthier housing (primarily rural not urban) and how to promote their construction. At that time, the British government, like the French, continued to be reluctant to place the responsibility for housing into the hands of the state. Both imagined that the market, if properly nudged, could be prompted to build better housing for the indigent and working poor. For that to occur, however, there had to be some kind of profit incentive. Roberts's book demonstrated that there were particular models and construction techniques that could be inexpensively built and be profitable.[29]

As an architect, Roberts was, not surprisingly, mainly interested in the structures themselves. When greenspace does appear in his designs, it is usually only as a shared, drying garden for the laundry of multiple families or an enclosed an eighth of an acre surrounding a single-family dwelling. However, to accompany his design for a lodging house intended to provide housing for young, unmarried men, Roberts wrote that, if an acre to an acre and a half could be set aside, it ought to be made into an "allotment garden" where a young man might spend his leisure hours "profitably" rather than "passing his evenings in a beer shop."[30] Years later, in 1863, the French reformer Alphonse Bertelé similarly recommended the inclusion of garden space in any plan for workers' housing, whether single or multi-unit. Unlike Roberts's discussion of housing for young, male, single workers, Bertelé's interest centered on the impact of accessible garden space for the working-class *père de famille* (head of household). "[I]n his moment of leisure," he contended, "the owner or renter could busy himself in the garden and thus abstain from going to the cabaret. It is, moreover, thrift for his household and recreation for his family, who can go there and breathe fresh air."[31] Gardens and clean fresh air in this sense would not only prevent a young man from wasting himself in the tavern and embarrassing his parents, they could also reach the all important père de famille whose conduct had a direct impact his family's future. The social significance that Roberts and Bertelé attached to greenspace emphasized gardens and gardening as an activity. Later works continued this way of thinking about gardens as an antidote to dissipation and expanded it to include public greenspaces that were not cultivated, such as parks and squares, particularly in working-class neighborhoods.

The concerted effort on the part of the French government to circulate Roberts's ideas speaks to the interest that the many shared with Louis-Napoleon in ways of creating healthy and affordable housing for the working class, but it also demonstrates that Bonaparte and the government of the Second Republic were reluctant to take on directly the responsibility of funding and building workers' housing. Indeed, housing remained a market where successive French governments were loath to intervene.[32] Rather, the responsibility fell to local officials to address the problem within their districts. They could, with book in hand and the sanction of the national government, encourage private investors or philanthropists and developers to bear the cost. This new approach departed from the position that Frégier and others had taken a decade earlier. Frégier had argued in favor of governmental responsibility for housing in his 1840 treatise, even if his language, ideology, and logic did not reflect the benefit such a plan would provide for those in need. The administration, he asserted, could do little to stop a poor worker from rendering himself senseless (with drink) in an effort to cope with the misery of his living situation; the administration had to "tolerate what it could not prevent," but that same toleration did not exclude building housing at its own cost intended for the those at the very "lowest rungs of society." Frégier continued by once again linking his argument to the dual benefits of crime reduction and salubrité. "The city, in light of the general interest, believes itself obligated to undertake the draining of sloppy water from a city like Paris by systems constructed at great expense; isn't there even greater reason to address another mess in the form of the miserable and vicious classes that swarm within the vast capital?"[33] Frégier's troubling analogy betrays his insensitive and dehumanizing attitude toward the poor, and his opinion had little to do with altruism; still his message and goal remained the same—the government must do something about the living conditions of the urban working poor because it has a deleterious affect on their lives, and by extension, on society.

Reformers, doctors, and administrators wrote feverishly of the consequences of unhealthy housing; the value of wide-open public ways; the physiological need for fresh, clean, well-circulated air; and the rights and responsibilities of private and public entities. As they tried to encourage good (and wealthy) men to act virtuously, some turned to what they saw as simple logic to address the problem. If one accepted that cities are inherently unhealthy, and that nature and the countryside are restorative and healthy, how could cities be made more like the salubrious countryside? The answer appeared quite simple—more of the natural world incorporated within the built environment would improve living conditions. Michel Lévy envisioned these kinds of alterations to the Paris streetscape and the "mephitic canals" that had worsened the effect of cholera and contributed to the unhealthy conditions in which the poor lived. The streets should have a row of trees planted between the road and the windows of the building, he argued in 1844. These plantings would "intercept the effluvium of the road, produce a particular freshness in summer and in altering the view, would have a favorable impact on one's morale."[34]

Horticulturalist Alphonse du Breuil extolled the value of trees in his 1848 textbook on the subject: "Let us remember as well that trees act powerfully on the health of mankind and animals, purifying the atmosphere and thereby making it more suited for respiration. . . . Thus, it is with good reason that one advises multiplying the plantings in and around large cities and habitations."[35] To buttress his own argument about salubrité in the urban environment, Tardieu in his *Dictionnaire d'hygiène publique* drew the reader's attention to the work of Michel Chevreul concerning the purification of cities. In a presentation made on 16 November 1846 at the Academy of Science in Paris, Chevreul had argued that three curative measures would reduce the amount of organic material in the atmosphere that was susceptible to becoming harmful: the enlargement of streets, the constant flushing of those streets with fresh water, and the planting of trees. "Yet, the planting of trees must be done with intelligence," Chevreul had cautioned, "with regard to the number, the distribution and even the choice of tree. It is indeed important that the roots be able to extend their reach enough to satisfy the needs of the types planted, without ever being exposed to infected soil where atmospheric oxygen cannot penetrate them."[36] The planting of trees along public streets emerged in the public discourse following the cholera epidemics as a potential multilayered resolution to the tensions among housing reform movements, public health policy, private interests, the reach of the state, and the responsibility of administrations. Introducing this idea in this way—couched in a scientific rationale, and tied to a social policy of amelioration of living conditions within Paris—linked the solution to positivism once again just as recommendations concerning interior air and light in domiciles had done.

This relationship between scientific observation and discovery, along with the adoption of those findings in urban planning and public health, became one of the defining characteristics of Adolphe Alphand's greenspace development plan. Édouard André, Alphand's assistant and principal landscape architect, wrote in 1879 concerning what he believed had distinguished the Second Empire park builders and landscape designers from the likes of André Le Nôtre, Capability Brown, and Jean-Marie Morel. "*L'art des jardin* exists," he affirmed, "but artists are lacking. It had its great epoch in a time where the poets, the writers, the premier painters of France and England did not scorn to be at once the faithful and the followers, or the lawmakers. Since that time it has degenerated into a trade, to an industry, with rare exceptions in between. But it seems today to be reborn out of the ashes, to be rid of empirical procedures, to show inspiration, and to call to its aid, the conquests of modern science."[37] From André's perspective, the nineteenth-century Parisian greenspace development program was progressive, modern, and rested on the *terra firma* of science. Armed with this abiding faith in science, he, along with a network of reformers, technocrats, chemists, doctors, civil engineers, and horticulturalists of the park service contributed to the inherent particularity of the Parisian park system, even as they endeavored to produce a set of universal principles and reliable practices of greenspace development that would address what they perceived to be the most pressing concerns of urban life.

Indeed, the work of esteemed biologists and chemists such as Michel Chevreul, Justus von Liebig, Jean-Baptiste Boussingault, and Georges Ville, who were primarily interested in developing better crop fertilizers, nevertheless helped to provide a compelling scientific justification for increasing plant life in cities that could accompany existing social arguments. These researchers understood that industrial and human waste was composed largely of nitrogen and nitrogen compounds such as ammonia. These constituted the main pollutants of water and air in large cities like Paris, making life for residents increasingly uncomfortable. Boussingault's experiments on precipitation in Paris during the 1850s showed that the rain, snow, fog, and dew in the city had levels of the nitrogen compounds such as ammonia and nitric acid three times higher than rural areas. This concentration, particularly in fog, "affected the lungs in a painful way" and adversely affected the quality of life for all city residents.[38] The fog was so thick with noxious particulates that on one evening in 1857 Boussingault recalled that "the vesicular vapor was so opaque that on the Blvd. Beaumarchais one need only stand just twenty paces from a gaslight to no longer be able to see the light."[39] In addition to what they understood about pollution, scientists were learning more about nitrogen as an essential plant nutrient. What they could not determine, however, was how precisely plants obtained nitrogen. If they could discover this process, science might provide a mutually beneficial solution to both the challenge of urban pollution and low agricultural yields.[40]

These goals invigorated international interest and early nineteenth-century research into the question. In the early decades of the 1800s, Swiss chemist Nicolas-Théodore de Saussure's experiments on plants refuted the previously held belief that plants could obtain nitrogen as a gas directly from ambient air. He argued instead that plants absorb nitrogen in the form of ammonia found in the soil, air, water, and manure. In 1838, Boussingault, a Frenchman, experimented on clover to rehabilitate the old theory. He wanted to determine whether plants could, indeed, absorb nitrogen directly from the air and if nitrogen levels could be augmented during the growth cycle. His experiments were flawed, however, in that he was unable to remove completely all organic and microbial material from his soil medium and the purified air he used (something Boussingault did not realize). This error produced inconsistent measurements. His experiments were, therefore, inconclusive on the precise question, but they led the scientist to postulate that perhaps nitrogen had entered the plants through particulates in the air. The experiments of German chemist Justus von Liebig in 1840 rejected the idea of the absorption of nitrogen from the air, either as a gas or in particulate form and upheld Saussure's earlier contention that ammonia was the source of nitrogen in plants.[41] In 1855, however, well-known chemist and philosopher Michel Chevreul of the Académie des sciences joined with a panel of other notable scholars (including Jean-Baptiste Dumas) to review the recently published work of a young Georges Ville. In the committee's report, they found that Ville's methodology had been sound and that he had indeed proven that the amount of nitrogen contained in plants did not come only from ammonia in the soil, but also from the air.[42] The work of these researchers was new and exciting and dovetailed perfectly with the positivist notion that science held the key to solving humanity's problems.

City planners and hygienists began to take note of this body of research and incorporated it into their own discussions of the value of trees in the urban environment and the rationale for park building. In making his case for planting more trees in the city, Alexandre Jouanet, an attaché to the *Service des Promenades et Plantations,* presented a complicated "scientific" calculation in his 1860 essay *Paris et ses plantations.* He first determined the number and variety of trees and their leaf size so as to arrive at an average leaf surface per average tree. This he doubled (accounting for both sides of the leaves) and multiplied by the total number of trees in Paris to calculate the total leaf surface area in the city. Divide this number, argued Jouanet, by the population of Paris and you arrive at the astonishing amount of 140 square meters of verdure per person, which would be each Parisian's own personal air filtration system. Science, he claimed, "proved that vegetation purifies the air," which is why some public administrations were engaged in "zealously planting trees within city limits."[43] Jean-Baptiste Fonssagrives, a noted doctor and professor of hygiene in Montpellier, agreed with Jouanet. In his work entitled *Hygiène et assainissement des villes,* written more than a decade later during the Third Republic, Fonssagrives criticized those who had disregarded the insights science provided. He singled out hygienist Dr. Jean-François Jeannel, a vocal opponent of anyone who, in Jeannel's view, exaggerated the properties of plants. In a presentation before the *Société de médecine de Bordeaux* in 1847, the hygienist had challenged the notion of trees as filtration systems, an orthodoxy attributed at the time to Chevreul, who believed trees had a favorable impact on salubrité.[44] According to Fonssagrives, Jeannel had gone too far in his effort to challenge Chevreul, writing with astonishment that, "after having denied trees the power of purification that tradition had attributed to them, he [Jeannel] does not hesitate to consider them as exercising an *unfavorable* influence."[45]

Fonssagrives dismissed the strength of Jeannel's critique even while granting that there were some small regional differences in the extent to which trees contributed to salubriousness in cities. For him, Chevreul's original ideas emerged from the debate unscathed since detractors like Jeannel had, in fact, done very little to supplant them.[46]

These kinds of sweeping statements about the benefits of trees in Paris no doubt frustrated Jeannel as he continued to challenge their veracity throughout his career. Some twenty-five years after his presentation in Bordeaux, Jeannel delivered a speech to the *Société d'acclimatation* in the Bois de Boulogne in 1872 in which he repeated his challenge to the orthodox view on air, health, and trees. "Gentlemen," he began, "I have no intention of discussing the embellishment or the pleasure that trees offer, as much as I would like to talk about light and sunshine. Neither do I wish to deny in any absolute manner the purification produced by these plantings, but I would like to examine it closely to determine, as much as is possible, the real value."[47] Science, he conceded, had indeed by now "proven" that trees purify the air.[48] Yet what, Jeannel wondered, was the appreciable atmospheric cleansing that occurred as a result of plant respiration; and was that a good argument for planting trees in the city? He also set forth a complicated formula measuring the rate of air exchange in trees; the various quantities of pollutants added to the air through burning wood, natural gas, and oil; and population figures for Paris. His conclusion was that it would require a forest seventy-five times the size of the city of Paris to cleanse the air sufficiently for the number of people who lived in the capital.[49] Finally, he challenged his audience to consider that, if trees cleanse the air so efficiently, why is there no measurable difference between the composition of the air in the winter as compared to the summer when trees are fully leafed out?[50] He concluded his speech by repeating his contention that not only had the purification of the urban atmosphere by trees been exaggerated, but that trees might make life in the city *less* salubrious. He argued that trees in Paris were planted far too close to buildings. They prevented sunlight and warmth from reaching into apartments and to the lower floors, and they increased the dampness of interiors both through transpiration and by simply holding precipitation so close to open windows.

Jeannel was not suggesting trees be removed, however, merely that their distance from buildings ought to be adjusted and standardized. None of the men had gotten it completely correct. Jouanet's figures were somewhat questionable, and his conceptualization of the organic function of trees now appears overly simplistic given all that scientists have learned since the nineteenth century. For his part, Fonssagrives may have dismissed Jeannel's critique too quickly, refusing to believe that poorly placed trees may have a negative impact on the urban environment. Finally, Jeannel's valid question concerning the *real value* of trees clearly ignored what they contributed in other ways to quality of life in the city. What is intriguing about these arguments and works is the way in which they demonstrate the kind of positivist landscape within which ideas on science, health, and urban planning intersected and were shared.

Against this backdrop of the research done in biology and chemistry and the many discussions among hygienists, Napoleon III expressed his own faith in the healthful benefits of greenspaces. The neighborhood squares, with their proximity to housing, were as important as the larger parks to the emperor. During one of their meetings regarding the renovation of Paris, Haussmann recounted that Napoleon III "ordered [him] to never miss any opportunity to create in all the arrondissements of Paris the placement of the greatest number of squares possible."[51] They were to be enclosed planted areas similar to those he had seen in London while in exile, but fully public in the French nation's capital. The emperor's insistence fit with Frégier's view that the cramped squalor in which poor and working-class populations lived was both a threat to public health and to public order.[52] Haussmann recalled that "the Emperor never lost sight of the necessity to build in parallel with the transformation of the Bois de Boulogne and Vincennes, in the center of Paris, smaller parks and squares, planted spaces distributed across the surface of the city, where the working classes could spend their break from work in a healthful way."[53]

Like Frégier, the emperor was also convinced of the necessity of moral order. He was certain of the positive "moral influence" these squares with green plantings could exercise on the masses. In 1866, Haussmann appeared to concur with the emperor's assessment. In his yearly report to the Municipal Council he concluded: "I shall not end this summary review of the most important public works projects in the city, without updating you gentlemen on those of the *public gardens*. Some have misunderstood these gracious creations, and see in them nothing but costly fantasies; but I believe that no one can now contest the positive effect and the happy influence that they exercise on the very morals of populations. We therefore need no longer justify the very considerable expense that we have not hesitated to make for the squares of the former

Paris and also for those that bring joy already to the newer neighborhoods of Batignolles, Belleville, Charonne, Montrouge, and Grenelle."[54] Writing in late 1880s, the former prefect indicated that had not, in truth, shared the emperor's sentiment. He had his doubts about the relationship between greenspace and moral order, calling it a "hope" and a "generous illusion" that had yet to prove itself, but he added that the "good effect [of parks and squares] on public health was incontestable."[55]

In the midst of these various discussions and debates, public greenspace emerged as a significant component of the redesign of Paris. Under the direction of Adolphe Alphand, the park service renovated previously built parks, constructed new squares and gardens, and managed and maintained all of those spaces. Existing squares, such as the Place Royale (Place des Vosges) among others, and gardens owned by the national government such as the Tuileries Gardens, the Luxembourg Gardens, and the Jardin des Plantes fell under the jurisdiction of the park service as well. In 1868, Chief Engineer Jean Darcel submitted a *Tableau Statistique* to Alphand that detailed the parks and squares created by the park service since 1856.[56] This early report shows somewhat more activity in the central and western areas of the city; however, the emperor's vision and Alphand's work at park construction and renovation continued well into the early decades of the Third Republic with some installations planned during the 1860s and executed in the 1870s like the Parc Montsouris, Square d'Anvers, Square Parmentier, and some originating during the Third Republic. In this light, the greenspace development plan was more comprehensive and expansive than other alterations to the cityscape at the time. Indeed, while street and housing construction clearly privileged wealthier sections of Paris, the same cannot be said of greenspace development, which was executed in a somewhat more balanced way. As historian Michel Carmona contended, when it came to parks and gardens, in the final analysis, the popular quartiers were not neglected. Indeed, two of the largest new parks (Buttes Chaumont and Montsouris) and many of the smaller gardens and squares were created in lower-income neighborhoods.

"In the new arrondissements," Carmona asserted, "as to the squares in the [bourgeois] west (Batignolles, Boulevard Victor, Square de Grenelle, Place de Commerce), there are those that correspond in the [working-class] north and east: the two squares at La Chapelle (18th), that of Belleville (19th), and that of La Réunion (20th)."[57] Adolphe Alphand was unequivocal in his estimation of the result of what he and his team had done vis-à-vis health and expanding access to greenspace. In *L'art des jardins* he wrote: "It is needless to insist on the hygienic and philanthropic importance of these plantings, 'as necessary to the habits of the adult population as to the health of children, that they must be able to go there as much as possible, near to their home, in a place where they might be secure, all so as to breath salubrious air.' The poorest neighborhoods, the most populous, have not lacked favor in this distribution of air, light and verdure; and it was only justice!"[58] The work of the Service des Promenades et Plantations—its duration, ethos, and attempt at uniform distribution of parks and squares—casts a different light on the renovations projects and suggests that greenspace development as part of a public health program, and the personal convictions of men like Alphand, who remained in key posts despite regime change, played a significant role during the transformation of Paris.

The fact that Alphand mentioned children's health in his favorable estimation of the value of the work the Service des Promenades et Plantations had undertaken is not surprising given that he was writing in the 1880s when pronatalist concerns and fears of depopulation permeated government programs and the discourse concerning public health,[59] but this was not the first time clean air and children's health had been linked to public greenspace. In 1827, the *Conseil de Salubrité de la Ville de Paris*, a precursor to the *conseils d'hygiènes*, recommended building a square in every neighborhood, "fenced in and planted with trees, in which children of all classes might . . . give themselves up to exercise appropriate to their age."[60] Michel Lévy likewise emphasized the importance of fresh air and outside exercise to the health of children in his 1844 treatise calling for improvement of workers' housing. He cited statistical evidence gathered by fellow reformer Louis Villermé which demonstrated that infant mortality rates were higher in the crowded neighborhoods of Paris than in those that were not as cramped.[61] He argued that children should be exposed regularly to clean air and sunlight in what, as Lévy reminded his readers, the well-known German physician and philosopher Christoph Hufeland had called a "bath of vivifying air."[62] Hufeland had authored a book in 1805 titled *Macrobiotik oder die Kunst das menschliche Leben zu verlängern*, in which he outlined practices he deemed necessary to live a longer and healthier life. The book was quite popular and influential, particularly among the mid-century hygienists and reformers. Lévy also referred to the work of Alfred Donné, whose 1842 manual, *Conseils aux mères sur la manière d'élever les enfans* [sic] *nouveau-nés*, contained what Lévy considered essential childrearing advice for new mothers. Donné instructed mothers to be certain that their children were regularly exposed to fresh air. It was not sufficient according to Donné for children to breathe fresh air while walking with their parents along the street, or shopping, or riding in a carriage; rather, he insisted that children should "*play* in the open air."[63] Lévy's use of Donné's ideas to support his argument for better workers' housing speaks to an early emphasis on children's health as part of a reform movement and the leverage it wielded.

The triad of children, fresh air, and exercise continued to appear in texts written by medical professionals. In the 1850s, Dr. Louis Cyprien Descieux published a series of textbooks to introduce hygiene into the primary and secondary school curriculums. His *Leçons d'hygiène à l'usage des enfants des écoles primaires,* written in 1858, was introduced into schools by the *Ministre de l'Instruction publique* and reprinted in at least eight editions through 1879.[64] In his book, Descieux addressed his young (presumably male) readers directly on topics related to nutrition and physical health. He urged them to be diligent in their studies but to take care of their physical being as well. "Play and work," he told them, "are as necessary as daily nourishment."[65] In the lesson on "Respiration," he described graphically the physiology of breathing so that children could understand the vital importance of exposure to air that is healthy and pure.[66] He outlined the various "poisons" potentially present in the air they breathe daily, particularly city air. Concerned that he had frightened them, he added, "I shall reassure you today by telling you that the leaves of trees and all the green parts of plants are able to absorb the unhealthy gas that constantly emanates from the chests of animals and men, as well as the gas produced by burning bodies [organic waste]; that the wind, rain, and thunderstorms purify the air by circulating and washing it, and by the happy mixing that occurs in the upper atmosphere which renders the air clean and healthy and fit for respiration."[67] Descieux's ideas echoed work being done in biology and chemistry, and he took as a certainty that which science was still debating.

In another work published in 1867, Descieux explained the value of exercise to healthy muscular development. He warned of the consequences of a lack of exercise: "Under the fatal influence of inaction, legs become debilitated, arms as well, and hands become clumsy."[68] Without exercise, he asserted, one seriously compromised one's health. One should carve out time during each day and devote it to some kind of moderate physical activity. Exercise taken in the open air, particularly a walk in the park, according to Descieux, had been incorrectly regarded as "a distraction, a pleasure," which work-obsessed people readily deprived themselves of, and in which lazy people refused to engage. He warned of the long-term consequences of a sedentary lifestyle. "Smaller children who are prevented from taking the quantity of exercise they need, remain weak, delicate, and often become infirm for the rest of their days."[69] Although education without time out for exercise might produce some of the brightest minds in France, he concluded, it also produced "feeble bodies, incapable of accessing that knowledge which had been acquired at the expense of health."[70] Descieux's connection between exercise and academic performance presaged much of the emphasis on the two evident in medical and educational discourse toward the end of the nineteenth century.

Jean-Baptiste Fonssagrives was likewise as interested in children's health as he was in trees and the salubrité of cities. He used language that tapped into increasing concerns about infant health and mortality rates.[71] In the late 1860s and early 1870s he began publishing a series of works about the physical education of children and domestic hygiene. In his 1869 book, *L'éducation physique des jeunes filles,* he wrote that "outside air was as indispensable [for children] as milk." He argued passionately that "[i]f a mother's breast [*sein*] was essential for a child, it is toward an equally robust udder [*mamelle*] that his/her lips must also turn for life and health: that of exterior nature, a fine wet-nurse as well, that provides air to the lungs and vigor to the limbs and color to the cheeks."[72] Fonssagrives's emphasis on maternal breastfeeding fit similar arguments concerning children's health and maternal responsibility that were being made by members of the Académie de médecine at the time—arguments that called for government intervention into family life to protect children's interest and the interests of society, paving the way for legislation in the early 1870s.[73]

In likening fresh air and nature to breast milk, however, Fonssagrives added access to open, natural spaces to the list of those things necessary for healthy development in children. Like Michel Lévy, Fonssagrives drew at length from Christoph Hufeland's popular 1805 *Macrobiotic,* writing that Hufeland had so well developed and stated the necessity of fresh, outside air that no one could better articulate it. Hufeland had argued that fresh air was an "essential nourishment," no less indispensable than drink or food.[74] It should be a "sacred and inviolable law" that mothers must take their children out into the open air daily; "enclosed air is a mortal poison for these little beings." As to where to take one's child for this daily exposure, Fonssagrives explained that Hufeland suggested "a garden, or an area covered with grass and trees; because it is only the air in these places that might be a true balsamic for the child, and one is often quite mistaken to believe that it is possible to procure the benefits of the open air by walking amidst filthy streets full of the emanations of our large cities."[75] Children's health and infant mortality, maternal responsibility, fresh air, and access to open greenspaces were all woven throughout Fonssagrives's work and in his recommendations for the future vigor of France.

Indeed, others in the 1860s made similar connections between the open air, gardens, children's health, and the future of the nation, reflecting concerns about maternity, the family, and depopulation in the decade before pronatalist legislation. The author Arsène Houssaye spoke of the imperial apartments at the Tuileries Palace after a portion of the garden had been set aside for the imperial family's personal use. In describing the emperor's and the prince imperial's ground-floor rooms, he called the emperor's apartments "severe" yet said of the young prince's rooms, which opened directly to the open space and air of the garden, that they "betray, quite naturally, play and study."[76] As the heir to the throne, Prince Eugène represented (for those loyal to the regime) the future of France. In 1867, that future seemed bright, secured after all by a fit and intelligent youth—a product of precisely the educational regimen balancing fitness and learning Descieux had prescribed. Houssaye's attention was equally drawn to the other children who played in the remaining public section of the garden. The Tuileries had traditionally been the haunt of the wealthy, but since the redesign of the Champs-Élysées, it had fallen somewhat out of fashion. Nevertheless, there remained a profusion of children in "swirling eddies," as Houssaye called them, conjuring up images of a fish nursery in a country brook. There, he wrote, one witnessed "the future of France at play with a hoop and stick."[77]

The Tuileries, like the Jardin du Luxembourg, had been designed in the formal French style centuries earlier when Le Nôtre designed gardens to convey the *gloire* of France embodied in the king. In mid-nineteenth-century France, the gloire of the nation increasingly lay in its children, and *they* now romped about these sculpted gardens of the *ancien régime,* just as they did in the neighborhood squares of a new, modern Paris. The Irish horticulturalist William Robinson analyzed the new parks and squares of Paris in his 1869 publication on the subject, which was revised and reprinted in numerous editions. In it, he recounted the sentiments of one of his colleagues, a Frenchman who lived in Paris at the time, and who, after having strolled through the Square du Temple one afternoon, related what he experienced to Robinson: "Some time ago, while walking through the Square du Temple, where hundreds of children were running and jumping and filling their lungs with the country air that has thus been brought into Paris, we could not help saying to ourselves that, strengthened and developed by continual exercise, these youngsters would one day form a true race of men, which would give the State excellent soldiers, good labourers for our farms, and strong artisans for our factories."[78] The sentiments expressed by Robinson's friend reflect the increasing fear of depopulation and the need for a strong military that gripped the nation in the days leading to the Franco-Prussian War and sparked new legislation in the wake of France's defeat. Still, there was more to it. Robinson's friend described a planted public square that had been constructed in a working-class neighborhood at the peak of greenspace development—one in which children played and exercised breathing fresh, clean air, and in so doing they were securing the economic, cultural, and political future of France. This close relationship between air, children's health, and greenspace which had been forged out of an earlier social reform movement and then imbued with concerns over depopulation and degeneration, now appeared fused to a new modern rationale and sensibility about public greenspace.

The many connections among Saint-Simonian philosophy, positivism, scientific experimentation, urban planning, and public health helped shape and define the Parisian greenspace development program and established a multifaceted argument for increasing the amount of planted spaces within the city. This is not to say that a garden aesthetic or *l'art des jardins* was not a priority in the work of the service, or that other social, political, and cultural factors had no bearing. Nor is it to privilege these connections over all others. Rather, acknowledging the role of philosophy, science, and public health in greenspace development adds an additional layer to our understanding of the program and the interest it generated outside of Paris. It suggests, as well, something about the larger intellectual and cultural currents of the time. In this way, throughout the decades of the mid-nineteenth century, access to clean, fresh air in Paris became closely tied to discussions of health, rights, and human necessity. The product of that linkage—public greenspace—engendered a sense of entitlement and collective proprietorship toward this particular kind of urban space, which differed from less *natural* spaces of the city. Thus, after their construction, parks and squares in Paris, came to be a nexus of community and group formation. Those multiple interests, imbued with this new sensibility, sometimes competed, making public greenspace a dynamic and contested space within the city.

GREENSPACE AS A WORKPLACE

••• •••

Being a park guard isn't just about watching out for misdeeds.
—AUGUSTE PISSOT, CONSERVATOR OF THE BOIS DE BOULOGNE, 1853

••• •••

Although national interests and health and welfare reforms of the mid-nineteenth century provided a powerful rationale for greenspace development, and although municipal structures provided a mechanism for implementing and sustaining that development, parks and squares in Paris had to be built and maintained by working men and women. These Parisians, whose livelihoods depended on greenspaces, occupied a social realm between that of the park designers, engineers, and administrators, and that of the park visitors. As such, they embodied something of each world. On the one hand, as employees or contractors, they were part of the larger mission of the *Service des Promenades et Plantations* to create fully public urban greenspaces that served all Parisians. On the other hand, they inhabited the parks in a way that administrators did not, interacting with the park users and like them shaping the space through their activities. They were the public face of the park service, and they constituted a bridge between the state and the individual. Whether park workers lived in the neighborhood or not, they became part of that community through their employment in the parks and squares, and some even formed their own communities within individual greenspaces. These men and women were an integral part of the operational and social landscape of the parks and squares. As such, they helped to define the nature of public greenspace and its place in the lives of city residents in ways more proximate to actual use than those of administrators and reformers.

Workers in the parks fell into two main categories: those who were employed by the Service des Promenades et Plantations, such as the park guards and construction laborers (*cantonniers*), and those whom the park service either contracted or licensed, such as the concessionaires operating restaurants, rides, refreshment stands, toy stalls, and other services, which the administration deemed necessary for the public's full enjoyment of the space. The park service charged the guards with the surveillance and management of greenspaces. Their primary task among many was to insure that no one damaged the hardscape (buildings, fountains, paths, and so forth) or the softscape (flora and fauna) of the park, and to see to it that the activities of one person or group did not impinge upon the enjoyment of the space by others. These two functions reflected the central concerns of the park service throughout the nineteenth century and informed nearly every decision that agency made. As a division of the service, the guard corps enjoyed the most liberty in the execution of their jobs and, among those at work in the parks and squares, had the greatest impact on the experience of the space. By contrast, the cantonniers had the greatest impact on the constructed space but had much less interaction with park goers. The concessionaires rounded out the spectrum of workers, ranging from poor workingwomen to entrepreneurs with some capital resources. They were independent contractors over whom the park service nevertheless exercised tight control. The concessionaires' primary interest was commerce. They sought profits, accepted financial liability, and had some power to make business choices as long as they did not countermand the aims of the park service. Women formed a significant portion of the workforce of concessionaires, as they owned and operated many of the smaller stalls, which in some squares constituted the only commerce.

This workforce helped to realize the goal of providing healthy play areas, aesthetic and comfortable parks, and improving the quality of life for Parisians. Yet their workplace was often fraught with tensions and social complexities. The public nature of greenspace and the establishment of parks and squares throughout the city meant that for guards, cantonniers, and concessionaires their work environments presented unique challenges that involved class, gender, community and occupational organization, social geography, and commerce. Comprehending greenspace as a workplace and exploring how these workers and the park service navigated and negotiated these tensions offers insight into the nature of these spaces and their place in the lives of citizens and communities in Paris.

With the formation of the Service des Promenades et Plantations, the city formalized and expanded the existing guard service in the Bois de Boulogne and elsewhere by creating the citywide *Garde du Service* in 1856. Prior to the Second Empire, gardens such as the Luxembourg Gardens and the Tuileries were state held and administered, thus policed by royal or national guards, while the rare, small, private squares, such as the Square de Vintimille, employed a security force paid for by local residents. By the middle of the Second Empire, however, the city administration reestablished its authority over those customarily semiprivate spaces, and the park service assumed responsibility for their management and policing. National properties were jointly administered, yet policed by the park service, which designated special units within the guard service for those locations. Although all of the guards of the park service were subject to prefectorial authority as part of the city administration, they were entirely separate from the municipal police. As greenspace within the city continued to grow, the guard service grew as well. Although its responsibilities and membership expanded, the structure of the guard service remained virtually unchanged well into the twentieth century.

The park service established the Garde du Service following a military organizational model. An *ordre de service* (organizational directive) in 1856 outlined the responsibilities of the guard service.[1] Chapter 3, article 15, of the ordre du service stated that guards were "subject to a military regimen. Each employee must in consequence, obey without dispute the orders of his superiors, seeking recourse [when necessary] only by way of the chain of command."[2] Officers in the Garde du Service held military ranks such as *garde général, commandant, adjutant, sous-adjutant, brigadier garde,* and *sous-brigadier garde.* Positions in the ranks of the guard service included *gardes portiers,* who were responsible for guarding the gates and outer spaces of the large parks; *gardes forestiers (à pied* [afoot] and *à cheval* [mounted]), responsible for the woods and interior spaces of those same parks; *gardes cantonniers,* who secured park construction sites; and *gardes ordinaires,* who looked after the smaller parks and squares throughout Paris.[3] Ties to the military went further still since the guards named to the park service had to have been (a) non-commissioned officers with fifteen years of military service, (b) decorated by the *Légion d'honneur* (as was Adolphe Alphand), or (c) have received a military medal of honor.[4] These hiring criteria resulted in a corps that was advanced in age, and one that would, presumably, command the respect of park goers and city residents. By virtue of its organization and recruitment, the garde provided a kind of semi-retirement for some members of the military.

Large parks such as the Bois de Boulogne and the Bois de Vincennes were divided into *cantons,* and guards were assigned to each zone. In 1856, the staff of the Bois de Boulogne comprised: two *brigadiers,* fourteen *gardes portiers,* four *gardes forestiers à cheval,* twenty *gardes forestiers à pied,* and six *gardes cantonniers.* With the exception of the gardes portiers, who were in residence by each gate, two or more guards were responsible for each area within the park, and covered the duties of absentee guards when necessary.[5] Smaller parks had far fewer guards, and squares generally had only one. Throughout the second half of the nineteenth century, the abundance of squares across the city, and a persistent lack of manpower (partially due to strict hiring criteria), meant that many of the guards were responsible for the surveillance of two or more local squares.[6] Thus, guard presence in the parks was often uneven. Work schedules and hours varied for all guards sometimes according to the needs of the park, sometimes according to the needs of the guard. For example, if the park service required more personnel for a race day, concert, or another well-attended event, additional guards were assigned and received a bonus for the added time.[7] If a guard needed to take personal time to bring a grandchild to school, or to appear in court perhaps, he could easily do so without pay but without penalty, as long as he gave advance notice.[8] In keeping with a military model, the guards wore uniforms reminiscent of those of the French Army.[9] Depending on his position, a guard might receive a ceremonial baton or a saber. Only the night guards in the Bois de Boulogne and Bois de Vincennes carried firearms since officials believed they were more likely to encounter dangerous criminals hiding from authorities in the darkness of the woods.[10] The men kept detailed logbooks, which their superior officers checked regularly. An attaché to the brigadier periodically inscribed ordres du jour or memoranda in each of the guards' logbooks so that directives from the administration were certain to reach all members of the guard service.[11] There was much about working as a park guard that was familiar and quite attractive to many retired military men.

In the Bois de Boulogne, Bois de Vincennes, Parc des Buttes Chaumont, and Parc Montsouris, some of the guard staff and their families lived within the boundaries of the park. The park service constructed housing for them and made additional provisions for the guards in residence. The houses were simple but comfortable with a foyer, a large dining room, a kitchen, one or two bedrooms, an attic, a basement, and an interior toilette.[12] These homes were not equipped with piped gas for interior lighting even when it was common throughout Paris because of the harm that gas leaching from underground conduits could cause to plant roots. Rather, specially equipped wagons carried the portable gas and filled lead-lined reservoirs in the basement of each building.[13] For heating and cooking purposes, the guards received regular, generous allotments of firewood.[14] Despite some of the benefits, guards and their families were subject to a number of restrictions when living in the parks. In an 1861 memo to the guards living in the Bois de Boulogne, M. Pissot, the head conservator of the park, accused the guards' wives of disregarding some of the rules. Some of the women who lived beside the *saut de loup* (a ha-ha or recessed garden wall) in the park, he told the men, had been tossing their wash water against the wall and killing all of the plants in the area. Some were even throwing it on the nearby lawn, destroying the grass. Lye and other harsh cleaning chemicals had caused the damage to the plant life. Several of the wives of the gate guards, he further pointed out, had been hanging laundry in the front of their housing, not in the rear. Park visitors entering and leaving the Bois de Boulogne could easily see the gardens in the front of the houses and, according to Pissot, laundry hanging there was "far from being proper."[15] Living in the larger parks for the families of the park service would have been both a pleasure and a bit of a challenge in light of the constant surveillance.

The interactions between the guards and other park service employees were closely scrutinized as well. Alphand worked to cultivate an *esprit de corps* and insisted that all those associated with the park service respect one another regardless of their position. In 1858, an incident occurred in the Bois de Boulogne that was a breach of this operational culture, and it drew uncharacteristic anger from the usually even-tempered Alphand. In June that year, one of the landscape crew at work in the park, a cantonnier, was "mistreated" while performing his job. It is unclear in the ordre du jour, which Alphand issued the following day, what precisely that mistreatment entailed, but it seems several guards in the vicinity had done nothing to intervene and help the cantonnier, which infuriated Alphand. "Four guards of the Bois," he wrote, "failed in their duty by not stopping several men who were mistreating a cantonnier. I hereby publicly censure the guard Guth who was the instigator of this deplorable act of cowardice. The present ordre du jour will be read to every guard and inscribed in their log books."[16] Ordres du jour were routinely written in the guards' logbooks, but rarely was there an explicit directive to read the order aloud to the men. Guth did not lose his position (his name appears in subsequent records) but Alphand had inflicted a most severe punishment nonetheless. The military culture of the guard and the men being all decorated veterans who prized highly their honor and bravery made being accused of cowardice and public shaming a significant reprimand.

The management of the park service expected that guards in the parks and squares would perform their duties well, function as a team, and act as ambassadors for the service. In his 1856 *ordre de service,* Alphand outlined some of the specific requirements of the position and the way in which the guards should perform their responsibilities. The order stated that guards were forbidden to enter cafés or cabarets while on duty, and that they should not involve themselves in lengthy conversations with visitors or neighborhood residents so that they would not become distracted. They were encouraged seek the help of other guards or cantonniers nearby to apprehend individuals and convey them to the local police when those individuals were in flagrant violation of the rules of the park.[17] Most important, they were to "show in the exercise of their duties, the greatest regard and politeness toward park visitors and avoid shouting or arguing."[18] As an ambassador of the city administration, a guard had to have the correct temperament when dealing with the public.

In 1867, Jean Darcel, one of Adolphe Alphand's chief engineers, received a promotion within the park service, which was quite likely in recognition of his skillful management of the massive construction of the Parc des Buttes Chaumont. He compiled an advice manual intended for the several subordinates who would be taking over his duties. Darcel advised them to remind the men that they were to serve the needs of the population and should always show appropriate deference. "Park personnel," he wrote, "must never lose sight of the fact that . . . their functions, like those of all administrations, have been created to satisfy the legitimate needs of the public. They must therefore permit all liberty in the parks compatible with the upkeep and conservation of that property. Let people choose the path they prefer without forcing them to go one way or another. They [guards] should limit themselves to preventing the degradations of lawns and flowers, the formation of rutted trails or paths on the lawns or through the forests. And they must prevent any impediments that trouble the park visitors."[19]

Darcel wanted those who managed the guards to reinforce the points that mattered most to the service: preventing damage and allowing free use of the space as long as that use did not hinder another's enjoyment of the same space. After having received word in 1874 that some of the guards were not behaving appropriately toward the public, Alphand issued a directive reiterating some key provisions of the rules of comportment he had established over a decade earlier in 1862. Article 20, he reminded them, stated that the "guards should adopt the utmost regard for and politeness toward the public in the performance of their duties, and avoid all debate. . . . Useless violence is strictly forbidden." Article 21 read, "Every time the guards shall perceive a contravention they should approach the responsible party to make clear to him the necessary observations, without shouting and violence."[20] For Alphand, it was not enough for the guards to simply prevent damage to the park space or apprehend those who violated the rules. To be fully effective, the guards had to command the public's respect. "The public expects to find in the guards of the park service an elite corps," he told them. "It must be constantly shown to them that this opinion is merited, and to do that, each guard must take to heart the desire to show that he knows how to mix a just severity with an extreme politesse."[21]

The guards' interactions with the public must have become problematic, indeed, by the time Alphand issued his directive in 1874 because, in that same year, Darcel addressed a circular to those in charge of the guards in the two largest parks. In it, he included an excerpt from another earlier memorandum in which Alphand had instructed the guards not to argue with the public when handing out a citation or making a *procès-verbal*. "It may well be," Alphand had written, "that the public is in the wrong, but even in the case where [their] remarks might be unjust and aggressive, the guards must never cease to be polite and give an example of calm. Justice will be much better served by an inquiry into the facts of the case contained in the guard's report."[22] Darcel's address to the administrators on this issue provides insight into the priorities and culture of the park service, some of the troublesome behavior of the guards, and their rapport with the public and other municipal agencies. His words are worth quoting at length:

Instead of protecting the park against degradation, preventing the formation of furrowed paths, politely inviting park goers who are walking on the turf beside the paths to return to the path or to go sit on the lawns off to the side, our guards—when they are not chatting among themselves or reading their newspapers—go searching in the woods for imaginary offenses. They see in every couple an outrage to public decency, in every woman a beggar or a streetwalker whom, without any proof, they conduct to the Commissioner of Police; once there, they are astonished to find that these functionaries pay no attention to their estimation of the situation or their complaints. They sometimes even behave inappropriately toward these magistrates.[23]

Excepting flagrant offenses that had been witnessed by other park goers, Darcel told the officials to instruct the guards to limit themselves to things such as conserving the flowerbeds and turf, assuring that traffic flowed, and seeing to it that dogs were on leashes and not bothering visitors. Furthermore, they should be "very reserved about seeking to quash scandals, particularly when those events were happening far from the main paths, except in the case of homosexual activity [*pédérastie*]."[24] Historian William Peniston argued that, although homosexuality was not illegal in France, police officials associated it with criminal activity and vast criminal networks, which justified, for them, aggressive and at times extralegal policing.[25] This helps explain, to a certain extent, Darcel's exception in his directive. "Quashing" such activity, disrupting it, and then reporting what the guard had seen would have allowed the city police to investigate the men and the location further in their own way, which generally involved stake outs, entrapment, sweeps, arrests, and interrogation.[26] Returning to Darcel's instructions, the general problem appeared to be one of balance—either the guards were excessive and overzealous in their surveillance, or they were completely lax and unaware.

As important as it was for the guards in the Bois de Boulogne and Vincennes to get along well with the public and police authorities, it was even more essential in the small neighborhood squares. The guards of the squares often lived in the neighborhoods in which they worked; if they did not, they spent a significant part of the day there and became a de facto part of that community. Thus, complaints about poor job performance on the part of guards in the squares generally arose out of the lack of rapport a guard had with residents in the *quartier*. In June of 1870, the guard at the Square des Invalides found himself in trouble with the office of the prefect of the Seine over an incident that occurred during one of his shifts. The Square des Invalides was one of two pocket parks, nestled into the corners on the north side of the veterans' hospital, the Hôtel des Invalides. Both squares were frequented by mothers and children from the neighborhood, who often had some connection to the military since the area was also home to a significant population of military personnel working in the War Ministry, the central military storehouse, and the École Militaire.[27] One summer afternoon in the square, the guard Moreaux had a brief conversation with a Mme. Brincourt, who, unbeknownst to the guard, was the wife of a brigadier general in the Imperial Guard. The substance of that exchange resulted in a letter from the general to the prefect in which the officer complained about the guard's behavior. The park service launched an investigation into what had occurred to upset Mme. Brincourt so, and Moreaux was called upon to explain himself.

In his interview, he remembered the day and the incident well. About 11:30 a.m., on the eighth of June, he recalled, he "found himself sharing a bench with a woman, her maid, and a young toddler."[28] The child was having a bit of a tantrum and letting out sharp screams. In an effort to be friendly toward the mother, Moreaux claimed, he leaned over and said to the child in a polite voice, "Now that is being a bit naughty there, little one."[29] He knew right away that the mother did not appreciate his intervention because she signaled her displeasure with her "dry and short tone."[30] Her reaction, he said, "rubbed him the wrong way," and he felt humiliated because "more than thirty times a day mothers, maids, and people with small children use the impression of a park guard in uniform to calm their children and bring them back to order."[31] He had only tried to help, he said. After having heard the guard's side, the official went to meet with General Brincourt. The general, in the meantime, had been conducting his own research on Moreaux and had discovered that he was a former noncommissioned officer in the Imperial Guard and that his record was excellent. Armed with new information, Brincourt changed his mind and apologized to the investigating official for having involved the prefect's office. He was, he said, in the process of writing a letter asking the prefect to forget the whole business because the general was now convinced that the guard had not been motivated by ill will.[32] The official of the park service thanked the general and assured him that Mme. Brincourt could "continue to visit the garden of the Invalides alone [without her husband] without the slightest bit of danger, and she would find there, if it be necessary, all the protection that is her due."³³ It is unclear in the record whether Moreaux received a different assignment to avoid further unpleasantness for both the guard and the general's wife. Still, relocation of a guard to another square was seen as a reasonable solution to these kinds of conflicts between neighborhood residents and park guards; termination almost never occurred.

Moreaux's encounter with Mme. Brincourt illustrates the potential for tense social interaction in the squares. The public squares were urban spaces where classes and genders could easily come into conflict. An employee of the service, simply going about his workday could, in a moment of intended helpfulness, insult a "social better" or offend a *mère de famille* (good mother) and potentially find himself in a great deal of trouble. A woman of position, on the other hand, might be forced to share the space with others not of her class, and not in her employ. With only her ladies' maid to accompany her, she might even endure a *remarque indiscrète* from a male stranger concerning her child's character. Social stratification in nineteenth-century Paris was often supported through spatial separations. Apartment buildings, particularly those in predominantly middle-class neighborhoods, reflected social delineations—horizontally by floors and vertically between the presentable facade and the hidden operations of domestics toward the courtyard.³⁴ Yet, when those interior private spaces emptied into the exterior public space, fewer formal separations existed, and none within neighborhood squares. A square could thus become a social minefield for everyone. The profusion of this kind of urban greenspace and the subtle conflicts to which it could give rise meant that these public spaces would have to be negotiated in ways that other urban spaces were not.

The public's critique of the guards' performance was sometimes a result of how well the guard got on with the residents of the quartier, not how poorly. In a letter signed only "Démophile," sent to the park service in the summer of 1873, a resident of the neighborhood of the Square Monge complained about several aspects of the care and management of that square, including the performance of the guard who worked there. "Démophile" wrote: "The guards of the square perform their service so poorly that many a time I have seen mobs of little urchins [*gamins*] rolling all over the grass and shredding to pieces the flowers and shrubbery there for more than two or three hours in the absence of the appointed agent who parks himself at the local wine shop at all hours of the day."³⁵ The park service manager, who wrote the report on the complaint, indicated that the damages described were not quite as bad as the letter writer had claimed. The park service, he noted, often had to replace plants in squares all over the city damaged by "malevolence" done after the close of day when it was "excessively difficult for the guards to exercise an efficacious surveillance."[36] As for the criticism of the guards, "it is sometimes merited," wrote the official, "but this shortcoming is a consequence of the situation, because these agents have nothing much to do from six o'clock in the morning until eleven at night. . . . The result is that they form bonds with the people who frequent the square where they work and thus become accustomed to taking refreshment with them."[37] Some guards, he added, had certainly had their assignments changed for this very reason, but those were really quite few in number. Returning to the complaint and one of the guards in question, the official wrote that the man did indeed have "the weakness of frequenting the wine shop."[38] But the guard had since changed his residence, and he was given a stern reprimand as well. Unlike the guard Moreaux at the Square des Invalides, who suffered from a lack of rapport (which was often true in cases of a socioeconomic disparity between guard and the neighborhood residents), the guard at Square Monge and others like him often had a relationship with his neighbors that was (as far as the park service was concerned) overly convivial. Thus, depending on the location of the square and the personality of the guard, the dynamic between guardian and park users could vary considerably.

The difference in scale between the large parks and the much smaller squares shaped how a guard might experience that greenspace as a workplace and his role in the space and in the neighborhood. During the Third Republic, with fears of degeneration informing increasingly pronatalist governmental policies, the smaller squares began to fulfill the earlier promise of providing children a healthy playspace, which was crucial to their health and development and to national concerns. Consequently, the squares became a veritable nursery in which France's past military heroes watched over its future citizens and security. The guards' rapport with the children in the square, whether avuncular or contentious, often determined his reception among the population in the surrounding quartier. This was true for the guard Millet, who worked in the Square de la Chapelle located at the intersection of the working-class Eighteenth, Nineteenth and Tenth arrondissements. On an August evening in 1889, Millet noticed a young boy, whom he recognized as eleven-year-old Jean Donis, striking other children who were younger than he was. The guard escorted the boy to the gate, twice ejecting him from the park, and twice young Jean returned and recommenced his bullying. The third time the guard apprehended the boy, Jean threatened the guard and used vulgar language at which Millet reported he clapped hands on the youth and took him to M. Douçot, the local police. The police sent for the boy's parents and, when they arrived at the station, in the presence of Millet, the officer gave the parents "a most severe reprimand" and told them they should "keep a closer eye on their child."[39] Yet, after having recognized the "honorability" of the parents, the police official and the guard released the boy to the couple's custody. At the Square de la Chapelle, the guard Millet functioned as the caretaker of the children in the square, protecting the weaker ones from being bullied, and meting out punishment on the bully. He knew the children in the quartier by name and, in bringing the boy to the police station, he publicly exposed M. et Mme. Donis as lacking in parental responsibility, which for Millet and the authorities was the greater offense in this instance.

Millet did not always trouble to involve the police in this way. At times, the guard bypassed municipal authorities altogether and, acting alone, confronted a parent directly concerning the behavior of children in the square. One afternoon, in the Square Jessaint (a greenspace across the street from the Square de la Chapelle, for which Millet was also responsible) he witnessed two children running and playing on an area of the lawn that had just been rolled and reseeded with an expensive, hardier strain of English grass. He knew the two were brother and sister, and although the boy escaped, he managed to catch the young girl, Marguerite Roberts. Millet knew that she lived at 83 rue Philippe de Girard and, on seeing the damage the two had caused, he "scooped up" Marguerite and carrying the girl made the one-kilometer trek up the street to her parents' home.[40] On reaching the home, Millet described to M. Fontaine, the children's stepfather, the damage they had caused, for which M. Fontaine promptly offered to pay. "I decided to eschew all procedure [a formal *procès-verbal*]," the guard wrote in his report, "and I invited M. Fontaine to put in writing his promise to pay for the damage, which he hurriedly did."[41] The expense of the damage was later determined by the head gardener Blandau to be "the cost of four days of one man's labor and the expense of six kilos of English grass seed" more than forty francs.[42] Here again, Millet knew the children in the neighborhood and where they lived. He did not hesitate to conduct Marguerite home and discuss the situation with her parents. M. Fontaine may well have been grateful that the police had not been involved since he would perhaps have received the same kind of tongue-lashing that M. and Mme. Donis received. Millet acted as more than just an agent of surveillance in the square. He held a position as a kind of social arbiter *and* caretaker in the neighborhood, and the community treated him as such. Neighborhood residents seemed to accept this role, perhaps in the understanding that, in these and other instances, the children were often alone in the park with only the guards and other parents to supervise them. Thus, the residents of the quartier participated in recasting the job of a park guard as something beyond surveillance alone.

This kind of interaction in the squares among guards, children, and parents could just as easily occur when the parent of the child was present. In 1888 at the Square Montholon in a more affluent part of town, the guard Grad was on duty when he noticed some children running and jumping on the benches. "I told them to stop" Grad wrote in his daily report, to which a father of one of the two, quite agitated that the guard had scolded his children, said to one of the youngsters, "Tell that one there [the guard] to take a hike when he says anything to you."[43] Grad told the father (whom he later learned was Léon Jourdan, who lived only a block from the square on rue Bleue) to just relax and stay calm. At which point, according to the guard, the gentleman threatened him, saying he would write to the prefect of the Seine about this incident. "By that time," the guard recounted, "a crowd had formed, and everyone was telling him [M. Jourdan] that he was very wrong to encourage his children to behave thusly."[44] Grad escorted the gentleman to the police commissioner, who said he would take care of the matter. At the end of the guard's daily report, the brigadier in charge of the area added a notation saying that M. Jourdan had come to see him afterward and insisted he had never said he would write to the prefect, rather that the guard Grad had "not understood his words."[45] The interaction between the two men that afternoon speaks again to the kinds of conflicts that could arise for a guard in the performance of his duties. It also exemplifies the way in which family life was often lived in public in the squares and how childrearing could be scrutinized, if not by the guards, then by one's neighbors. In whatever part of the city, the squares were centers of urban villages in which the community often exercised some level of social control.

Being assigned to a small square where children played in significant numbers could also be a problem for a guard if he did not possess the necessary temperament, or if the children were particularly high-spirited. If the residents of the quartier perceived that a guard treated the children overly harshly, he could not expect the support of the community. In 1892, the mayor of the Fourth Arrondissement received a letter of complaint "on behalf of a large number of families" who lived in the area of the Place des Vosges concerning the guard on duty there. The spokesman for the group, the director of the École Communale des Garçons, which was located across from the square, wrote that the guard "prevents students from crossing the square, he insults them and strikes them with the utmost brutality, and he pursues them all the way to, and *into,* our school, causing scandals everyday."[46] Many schools used the greenspace as their recess space, the director said; however, this guard believed that all of the children who caused trouble in the square attended the École Communale, which was, in the guard's opinion, "full of nothing but hoodlums."[47]

While watching the activity in the Place de Vosges one day from his office window, the director observed the guard breaking up a snowball fight. Some of the boys involved went off in one direction, and some made their way to the school. The guard chased this group of boys, swinging his cane at them and calling them "hoodlums" all the way into the school. "Witness to this act of brutality," the director wrote, "I approached the guard in the corridor and I told him that I was going to inform the Administration of his conduct."[48] The guard responded "insolently," the director claimed; he threatened the schoolmaster, saying that he (the guard) would file a complaint on the director; "he left and continued to create a stir among the passersby inveighing against everyone.... Today [our students] have renewed dread of the guardian who is supposed to protect them."[49] The schoolmaster asked the mayor to press for the removal of this "brutal man, who causes a scandal everyday at our doorstep, and contributes to giving us a bad reputation throughout the *quartier* that we do not deserve."[50] Many Parisians perceived of squares as safe havens for children in their neighborhoods. The Place des Vosges, surrounded as it was by schools, was also a means by which the schoolboys could take some physical activity in the midst of their studies, as Dr. Louis Cyprien Descieux had recommended in his treatises on health and education in the 1860s. If a guard's behavior threatened that nature and function of the greenspace, members of the community took action to rectify the situation. The mayor forwarded the letter to the park service, which initiated an inquiry into the charges. There is little doubt that the boys were quite rambunctious during their recess breaks; in fact, the director of the school indicated that there were often "*batailles*" in the square between his boys and boys from the nearby École des Frères. Nevertheless, the official complaint addressed to the park service from the local school board indicated that their investigation had confirmed that "the attitude of the guard toward the public school children leaves much to be desired."[51] That official recommended that the park service "take the necessary measures to avoid a repeat of such conspicuous occurrences."[52] Unfortunately, the report does not indicate what final action was taken against the guard, although in similar cases involving verifiable complaints, guards were given another assignment.[53]

Guards in the squares often found that their job performance came under intense scrutiny by the public, whereas the cantonniers drew much less attention. The cantonniers occupied the lowest position in the hierarchy of the Service des Promenades et Plantations, but the job was attractive because of the steady employment it offered. There is much less information available in the historical record about the day-to-day work habits and environment of these men since they did not record daily activities in logbooks as the guards did (most of the cantonniers were in fact illiterate, as noted in park service personnel records), and they appear much less often in petitions or complaints made by park goers. Still, what is known provides additional insight into the nature of greenspaces in Paris. Although the cantonniers were part of the architectural division of the park service, that did not limit their activities to the hardscape constructions in the parks and squares. They also worked on the landscape crews that transported, planted, and maintained vegetation. Like the Garde du Service, the park service recruited the cantonniers from the ranks of the military and navy. Many had served as infantrymen, sailors, or for the army corps of engineers during the Second Empire's military campaigns (such as the Crimean War and the Italian campaign). The park service required that they have a clean military and criminal record.[54]

The cantonniers were organized by skill levels designated as classes. The starting monthly salary for chief cantonniers was 95 francs per month; for first-class cantonniers, 90 francs; and for second-class cantonniers, 85 francs.[55] Promotions and salary could increase a worker's salary to 125 francs per month, although there is no indication as to what constituted a workweek. The cantonniers were organized into work crews under a *conducteur,* an on-site engineer or gardener who managed the work. When the crew received a project, they met at a location determined by the crew foreman, were informed of their assignment, and received tools and uniforms for which up to four francs per month could be deducted from their pay.⁵⁶ The amount of time a cantonnier might work in a particular area depended on the project—squares and parks could sometimes take years to construct or renovate, requiring varying levels of manpower throughout the phases of installation. By the middle of the Second Empire, work on public greenspaces in the city had increased so much that the process of assembling and equipping crews had to be formalized. As of 1863, unless otherwise specified, all crews were instructed to assemble at the park service's central warehouse located near the *Panorama des Champs-Élysées* between 8 a.m. and 9 a.m.⁵⁷

Like other groups of workers during the Second Empire, the cantonniers formed a mutual aid society, known as the *Société de Secours Mutuels dite du Boulogne*. Initially organized for workers in the Bois de Boulogne, the society eventually served all cantonniers of the service. Mutual aid societies provided for such things as the material needs of injured workers and support to widows and families. These organized workers' societies were deeply troubling to successive governments during the first half of the nineteenth century, and were a source of particular concern for the Second Republic given the role many believed "socialist" agitators had played in the National Workshops and the revolution of 1848. On March 6, 1852, however, then-president Louis Napoleon legalized these societies, whether by way of political expediency or from an earnest belief in some social reform. Later, as emperor, Napoleon III made regular contributions to their funding.⁵⁸ Personnel records for the cantonniers report lump-sum payments to widows from the Société de Secours Mutuels dite du Boulogne in amounts ranging from 100 francs to 250 francs. The société could even help with medical costs if necessary. Jean Chareille's personnel file indicates that he became ill shortly before his death in 1878, and he petitioned the mutual aid society for help with the cost of his medicines. The société instructed Chareille to submit his prescriptions and be certain that he had them filled by a pharmacist whose prices the society had approved.⁵⁹ The laborer was then reimbursed for the cost of the medicines. Chareille did not survive his illness, and his widow received the customary assistance from the society of 100 francs.

Work as a cantonnier was steady, and the men generally kept their jobs unless they had received multiple reprimands for such things as being late, loafing, or drunkenness; or, if they had actively supported the Commune of

1871.⁶⁰ As a graduate of the *École des Ponts et Chaussées* with its military culture and ties to civil service, Alphand was deeply committed to order, hierarchy, and a lifetime of service to the nation regardless of the regime. With intimate knowledge of the city's infrastructure, he had played an important role in the army during the siege and as its troops retook the city from the Communards. Many in the military at the time viewed the National Guard as an ill-disciplined rabble lacking all order and respect for authority. Thus, Alphand would not tolerate anyone who had actively supported the Commune in the ranks of the park service.⁶¹ Isolated minor offenses were generally recorded in the personnel file but posed no immediate threat to the individual worker's employment. One laborer, Louis Flammant, reported to work one day in 1886 so inebriated that "the *cantonnier* Carton was obliged to carry him back to his home."⁶² He received a written reprimand and reported to work the very next day. On July 11, 1868, Jean Jouas filed a complaint with the park service, saying that his foreman, M. Labiche, accosted him as he was exiting a wine tavern while on his shift. Jouas claimed he had only entered for a second with two of his friends. His boss, he said, had "spun him around and insulted him." For his part, Labiche reported that it was in fact the second time that day he had seen Jouas coming out of the wine shop. The foreman told him to "get to work," to which Jouas responded "insolently," and the two argued back and forth. Later that day, according to Labiche, Jouas gave him a "coup de pied" (kick).⁶³ Despite this somewhat amusing scene of insubordination, one that could easily have resulted in dismissal, Jouas kept his job with the park service. There is no indication whether he remained on Labiche's crew. Based on various similar entries in personnel records made by their superiors, the cantonniers appear to have been a hardworking, hard-drinking, cantankerous group that, nevertheless, constituted the critical construction muscle of the park service. They may not have interacted as directly or frequently with the public as did the guards; however, they too occupied the space between city and parkgoer as their labor built and maintained the thousands of acres of public greenspace transfiguring the cityscape.

The many concessionaires who populated the parks and squares similarly helped to shape the space and affect the experience of park goers. They were primarily small businessmen and women to whom the park service granted permission to operate businesses or provide services within the parks and squares. Their operations were strictly controlled by the city, but they were not city employees. Many became part of the social fabric of the quartier and, as was true with the guards, they might also come into conflict with some residents as they attempted to go about the business of making a living within these new urban greenspaces. The park service determined what concessions to allow in public greenspaces and in which locations. The process for obtaining a concession was quite simple. Businesspersons submitted a request for the right to conduct business in the park or square. In their petition, they described the enterprise and explained any additional relevant information such as their own experience, or the need and value of such a concession in the park. The park service exercised full control over commerce in the parks and determined what types of concessions it would allow, basing its decision on four general criteria: the potential damage to plant or animal life; whether the concession would prevent others from using the space; the existence of the same or a sufficient number of such offerings in the park; and the usefulness and appropriateness of the offering to the park goers and neighborhood.[64]

During the second half of the nineteenth century, amusements in the parks included children's entertainment such as merry-go-rounds, swings, seesaws, goat and donkey cart rides, marionette theaters, toy vendors, and candy stands. To accommodate adults there were restaurants; ice skate rentals and lessons; telescopes; refreshment stands; chair rentals (when benches were occupied); waffle stands; and horse, boat, and carriage rentals. Toward the turn of the century, bicycle rentals, mobile photographers, and stereoscopes began to appear in the greenspaces as well, reflecting some of the mechanical and technological advances of the late nineteenth century. If the park service approved a concession, the owner was required to pay a yearly fee for the license, and agreed to any stipulations the service placed on the operation, which could include fare restrictions and price caps. Some businesses moved into existing structures built by the park service—such as restaurants, refreshment stands, or the ubiquitous onion-domed multipurpose kiosk. In the case of the larger structures such as the restaurants, owners lived in the buildings, paid rent, and became part of the community that lived in the parks.[65] Other concessions were mobile ones such as cart rides, carriage and horse rentals, and the like. These concessionaires carried the necessary tools of their trade in and out of the park each day. There were strict rules about the specific area within which each could work, and a breach of those rules could result in a revocation of the license.[66] Mobile concessions that did not require equipment such as skating instruction became more strictly regulated during the Third Republic, when official identification badges were required. In the 1850s and 1860s, these kinds of mobile businesses were often just granted a *titre de tolérance,* which was less formal and regulated than an actual concession requiring a contractual agreement, and did not carry a yearly fee.[67]

Concessionaires within the parks and squares were accustomed to a wide array of rules and constraints. For example, in a memo dated 1 July 1853, Haussmann defined the conditions under which carriage and horse rentals could operate in—or more precisely, just outside the Bois de Boulogne. These conditions were eventually extended to the Bois de Vincennes, Parc Monceau, and Parc des Buttes Chaumont, the only parks that permitted vehicular traffic. The order read: "Rental stations will be placed in such a manner as to present neither encumbrance nor obstacle to access to the gates of the parks."[68] The carriages and horses were to cluster outside whichever gate they had been assigned, and they could not change locations at will. The park service limited the number of horses and vehicles each lessor could have in operation at any given time. All animals were to be in good health, and their tack was to be well maintained. Neither the concessionaires nor their assistants were allowed to hawk their services within the bounds of the park. Finally, the order forbade the rental of animals or vehicles to persons deemed "incapable of driving" such as children, drunken men, or women.[69] There is no indication in the *ordre du jour* whether livery companies paid a yearly fee for this authorization, but it is likely since such concession fees were a primary means of generating revenue to help pay for maintenance (in this case to offset the cost of maintaining carriage paths). The restrictions and regulation of this industry gave the city some control over the flow of traffic in the park, helping to reduce the incidents of collision, protect the safety of pedestrians, and preserve some of the tranquility of the greenspace.

The park service similarly tightly controlled the operation of restaurant concessions. In 1876, the manager of the Pavillon Puebla in the Parc des Buttes Chaumont petitioned for the right to install a piano so that restaurant customers could engage in an occasional impromptu sing-along, if they wished.[70] The permission was granted; however, the park service stipulated that the piano could be for the use of patrons only, and thus no "paid" entertainment would be allowed. There is no mention in the report as to why the park service made such a distinction; it certainly wished to limit the extent of commerce in the park, and to hold prices down particularly in Buttes Chaumont (both explicit concerns that appear in many of the decisions regarding concessions in the parks). The stipulation also reflects the longstanding concerns of the city police who monitored theaters throughout Paris to uphold decency standards and, during the Second Empire, prevent the potential for political agitation rife in the café concerts.[71] An addendum to the authorization indicates that this was a new kind of request, so permission had been granted on a trial basis only. The addition clarifies that this is not a formal, contractual concession; rather it was a simple *tolérance* and, since the park service could not predict what the impact might be, the moment this new activity posed any inconvenience to other park visitors, it would be revoked.[72] With a proliferation of public greenspace throughout the capital, and little experience managing or using it, the park service and the public learned together what each required of the space, as well as what could or would not work.

In addition to operational constraints, restaurateurs within the park could not make any physical alterations to their establishments without prior approval from the park service. They did not own the property; they leased it, even if they were in operation for many years. In 1885, M. Bouquet, who had maintained a restaurant concession in the Parc des Buttes Chaumont for nearly twenty years, petitioned the park service for the right to open a window in the kitchen of his establishment. The air and light were insufficient, he claimed, and he offered to pay for the work. Inspectors from the park service visited the site and reported that the kitchen was indeed "uninhabitable and unhealthy."[73] The report recommended approval of a window on the side of the building, as long as it was hidden from view by the bushes. In addition, while the work was underway, there could be no alteration to any of the plants around the building or along the approach to it.[74] The park service was not always so amenable to alterations made by or for concessionaires, particularly when it came to the plant life in the park. Years earlier, when a certain M. Javelot assumed operation of his concession at the Pavillon du Chemin de Fer in the Parc des Buttes Chaumont, he asked the park service to chop down three large trees that stood next to the building. He claimed that they increased the dampness of the area, and workers had to wipe the tables beneath the trees daily and even several times after a rainfall since the leaves retained the precipitation. The park service official answering the request stated that those trees had been planted at considerable cost to the city and at the request of M. Bridoux, the former concessionaire, who wanted shade under which his customers could sit enjoying their food and drink. "As we can not plant and replant trees at the whim of concessionaires," the official wrote, "we have decided to decline the request."[75] The Service des Promenades et Plantations was more likely to alter the park space at the request of citizens and park goers than at the request of concessionaires hoping to improve the profitability of their operations. To the park service, the concessionaire was not the intended beneficiary of park space, the park visitor was. Thus concessionaires, like guards and laborers, occupied that middle ground between the park developers and the park visitors.

The Service des Promenades et Plantations was constantly concerned about the impact of the structure of the concessions on the physical space of the park, as well as the safety and experience of the park visitor. In the 1868 statute that established the first two restaurants in the Parc des Buttes Chaumont, the city spelled out its conditions on M.M. Manceau and Bridoux's operations. There could be no signage in front of the structures that would mar the natural beauty of the park, and interior lighting would be dictated by the administration. Animals of all kinds were strictly forbidden on the site (no dogs running about or barking), and events such as large parties or balls had to receive prior approval. Wastewater and garbage had to be disposed of as directed by the park service since harsh chemical cleaning agents made wastewater particularly toxic to plant life, and garbage could rot and attract vermin if not disposed of properly. All employees required city approval and, if necessary, the concessionaire agreed to fire any employee the park service no longer wanted working within the limits of the park. Again, no structural changes could be made to the building whatsoever; rather, any desired changes would first have to be approved by the park service, and then built in accordance with its design specifications.[76] In short, concessionaires had much less say about the operation of their businesses than they might have had outside of the boundaries of the park. Still, these enterprises offered entrepreneurs the dual incentives of the potential for a modest profit and the chance to live comfortably in a country setting within the city.

Finally, the statute expressly forbade damage to the trees: "M.M. Manceau et Bridoux shall not touch any trees in the area of their concessions. They must abstain from any activity that could destroy or damage them. In the event of such damage, they will pay a fine to the city equal to ten times the value of each tree destroyed or damaged, and they will be subject to all penalties in the law governing material loss. They will be subject to replacement costs every time the Administration deems it necessary."[77] No other stipulation in the agreement between the city and the concessionaires carried a similar fine and threat of legal action. Concessionaires could live and work in the park but, in keeping with the park service's nearly singular obsession, they could not cause any damage to the property. This preoccupation with damage to plant life such as trees, flowers, and lawns was not due to any horticultural zeal, but reflected the great expense and difficulty of engineering these ostensibly natural spaces within the unnatural environment of the city, and the focus of the agency (composed of problem-solving engineers) on functionality and working systems. A relentless pragmatism, born out of the positivism of the mid-century, informed all of the decisions of the park service.

The Service des Promenades et Plantations guarded the spaces of the parks jealously, and its concern over the effect a structure might have on the interior spaces, the aesthetic or experience of it, translated to much smaller concessions, both before and after completion of the park. In 1865, while the Parc des Buttes Chaumont was under construction, the park service denied a petition by Mme. Laurens to construct a wooden stall within the boundaries of the park for the purposes of selling beer in the morning to the men working on the installation. The engineer who handled the petition wrote that he saw no problem with the sale of beer to workers, but he declined the request because he could not authorize a "permanent installation" such as the "shack" Mme. Laurens proposed to build.[78] Mme. Laurens could sell beer, coffee, or even *eau de vie* to the workmen in the park if she chose, but only if she circulated using baskets, as was the case at other jobsites throughout the city. Here, it was not the nature of the concession that disqualified it, but rather the fact that it did not fit the overall plan of the park, or even of the worksite.

For the park-building engineers, everything in the space had to function according to a conceptualized design and program, at least initially. That which might impede any particular function could not be allowed, while that which would have little if any negative impact on the space might easily be permitted. This was evident in the case of M. Monsus, a photographer who requested the right to move about the Parc Montsouris and photograph park visitors, selling them a souvenir postcard as a memento for a small fee. The manager who reviewed the request said he believed that allowing M. Monsus to photograph park goers presented "no inconvenience" to the park space.

Monsus's operation was a mobile one, and his equipment consisted of no more than a folding chair and a canvas backdrop, which occupied a one-by-two-meter space. The park service thus granted Monsus's request with some stipulations, such as requiring that the photographer carry all of his equipment in and out of the park, that his set-up never exceed three square meters, that he pay a ten-franc annual fee, and that he confine his operation to specific locations deemed appropriate by the park service.[79] In this way the Service des Promenades et Plantations maintained its strict control over all concessions in the park, including mobile concessions, insuring that no operations upset the functionality and design of the greenspace.

If the park service did grant permission for a concession that required a more permanent structure within the greenspace, the cost and construction was sometimes the responsibility of the concessionaire, even though the structure would be subject to the specifications of the service. In 1888, M. Bouquet again petitioned the park service; this time it was for permission to establish another business—a small refreshment stand by the lakeside. The park service approved Bouquet's request, but required that the stand be diminutive (no larger than 3.75 by 2.50 meters), entirely built of wood, so as not to spoil the overall perspective of the park, and not be situated by the lakeside, but rather in a location the park service would designate.[80] The spot decided upon was set back away from the lake on the incline of a nearby slope and nestled in a clump of trees. The report stated that the site was chosen because it would be sufficiently hidden from view for those entering the park at the Laumière gate.[81] Bouquet's refreshment stand might have served the needs of thirsty park visitors by the lake, but the park service had to balance that benefit with the impact it could have on the overall aesthetic of the park and the experience of others entering at another point.

As much as concessionaires and their commerce were strictly controlled by the Service des Promenades et Plantations, and while the administrative agency privileged a park visitor's needs over those of the concessionaires, it was not a cold, bureaucratic machine. As restrictive as the park service could be at times, it could also demonstrate a keen sensitivity to the challenges facing its "partners"—the concessionaires. This was particularly evident in exchanges between the city and the concessionaires of the Parc des Buttes Chaumont in the aftermath of the siege and the Commune. In August of 1871, the manager of the Pavillon Puebla, M. Bouquet, contacted the park service requesting that he be forgiven a portion of his rent and to have his lease prorogued, its terms frozen for a period of five years. Bouquet claimed that he had suffered significant financial loss from the closure of the park in August of 1870. At the start of the war, and throughout the days of the Commune, the park service handed some of the parks and squares in Paris over to the military for use as encampments, staging areas, or for the strategic placement of artillery.[82] Bouquet pointed out that his establishment had been ruined by the subsequent bombardment. Moreover, the structure and his possessions suffered considerable damage when the Communards took control of the park, making a last stand on the butte there until heavy shelling by government troops on the heights at Montmartre forced them to abandon the position.

Bouquet asked the city to forgive his rent for the period from August 1870 to January 1872 since he had no effective use of the space during that time; and he asked that the city suspend his lease for five years, allowing him to recover financially, after which all of the original terms would be reinstated. The park conservator, M. d'Arboussier, reported the request to chief engineer Jean Darcel. He wrote that he believed the rent *should* be forgiven, but only from 1 September 1870 (the day the park was closed) to 15 August 1871, the date Bouquet could have used his restaurant again. As for damages suffered as a result of the bombardment, the official indicated that this was the responsibility of the War Administration, not the Service des Promenades et Plantations. Finally, with regard to the prorogation of Bouquet's lease, d'Arboussier suggested approving the request, saying it might be just as well for all concerned. "It is necessary to recognize," he wrote to Darcel, "the situation over these past two years has been unfavorable for the *quartier* of the Buttes Chaumont, and it is feared that it will be thus for many years to come."[83] Darcel approved both the forgiveness of the rent and the prorogation of the lease. Bouquet took up his concession again as he intended and moved back into the park in 1879. He remained there for several years, and his name appeared on an 1889 list of fishing licensees in the park.[84] The park service worked with Bouquet when times had taken an unfavorable turn, further deepening the relationship between the administration and the concessionaire and providing for increased continuity in the function of that concession.

Bouquet was not the only businessman in the park who experienced the tribulations of the siege and the Commune. M. Bridoux, concessionaire at the Pavillon du Chemin de Fer, suffered as well. Bridoux's establishment was not subjected to as much physical damage as Bouquet's, yet in 1872 he similarly contacted the park service to request forgiveness of a portion of his rent for the last quarter of that year, and asked that he be released from his lease as of January 1873. His wife, he indicated, had suffered something of a nervous breakdown on the heels of the "emotional distress she experienced when the *fédérés* occupied the park."[85] Not only had the park been occupied, some three hundred corpses had been thrown in the artificial lake by the army after taking the ground from the Communards. The bloated, putrid bodies remained for days until they were pulled from the water and burned.[86] According to Bridoux, the experience of the Commune had left his wife "extremely sick."[87] Since he had not received the interest on his initial deposit for the rental, he wished to apply that to his last rent payment. Darcel acceded to this request as well. Unlike Bouquet, Bridoux and his wife left the park, never to return. Acting on behalf of the service, Darcel and d'Arboussier showed flexibility in their dealings with these concessionaires, demonstrating an understanding of the unique situation and the personal losses each had suffered.

As in the case of Bouquet and Bridoux, the park service could at times behave with some sensitivity to the financial and family situations of those seeking concessions; it weighed those concerns carefully in their decisions to grant or deny requests. Smaller concessions could function as a kind of social safety net, a welfare plan, for those the park service deemed most in need and most worthy. Those who received these smaller concessions were former employees of the park service, invalids such as wounded soldiers, or widows of employees whose husbands had served the city long and well. In the absence of a real pension plan, the park service could still provide some kind of income for these individuals. In 1890, a veteran of the Franco-Prussian war, M. Belotte, asked to be allowed to set up a flower stand in the Square Avenue de la République (now Square Samuel de Champlain) beside the east entrance of Père Lachaise Cemetery. The appropriateness of a flower stand at the gate of the cemetery certainly played into the decision to allow such a concession, but Belotte's history and financial woes were a much more significant factor. Belotte had been a gardener prior to the war, but a gunshot to the leg crippled him and made it impossible for him to practice his trade any longer. The park service official who reviewed the case noted that M. Belotte supported a wife and three daughters on only a small military discharge bonus of 190 francs per year. He wrote that his office had received glowing reports of Belotte's character and that he believed the park service should grant the request based on that input, and the pressing financial need of the veteran.[88]

Belotte's case notwithstanding, small concessions and stalls were more often than not granted to women who had some kind of familial connection to the park service. The agency's practice was to categorize the various concessions and designate those particularly suited to serve this social welfare purpose. In a 1902 report concerning a dispute over the right to operate a kiosk in the Place des Vosges after the previous concessionaire had become too ill to continue, a park service official reminded his colleague in an addendum that the concession in question, "being one of favor should be returned to the category of those concessions reserved for former agents and workers of *the Service des Promenades* [sic] who find themselves in need, and to widows or beneficiaries of those agents or workers."[89] Indeed, although men owned and operated the larger concessions (such as restaurants, or carriage and horse rentals), widows of park service employees were frequently the owners and operators of refreshment stands, toy stalls, toilette and seating concessions.[90] Thus, from the earliest creation of the many parks during the Second Empire through the first decades of the twentieth century, commerce in the neighborhood squares and pocket parks was gendered female, as in many of the markets of the city.[91] The fact that mothers and their children frequented the smaller neighborhood squares is one possible explanation for the distinctions in ownership of businesses in the parks. Although the larger parks of the Bois de Vincennes and Boulogne contained small concessions owned and run by women, the larger and more lucrative concessions in those parks continued to be operated exclusively by men, or by men and their wives.

Since women operated most of the smaller concessions in the squares, their presence and involvement in commercial activity across the city often led to conflicts between those working businesswomen and wealthy, bourgeois neighborhood residents or local businesses. A common job reserved for workingwomen in the greenspaces of Paris was that of a *receveuse*. A receveuse was a woman who had obtained a concession from the Service des Promenades et Plantations to rent a specified number of wooden or wrought iron chairs to park goers in a designated area of the park or square for a small fee. Although the park service tried in earnest to provide sufficient seating in the form of fixed benches, there were often simply not enough to accommodate the large numbers of park goers. If the benches were full, visitors could bring their own folding chair into the park if they wished, or they could take advantage of the services of the receveuse. When these workingwomen received a concession in squares in the affluent west side of Paris, they often found their job complicated by a disdainful attitude on the part of neighborhood residents. They were accused of any number of offenses, from rudeness to prostitution. These complaints usually involved a report of behavior the complainant considered inappropriate, or some kind of confrontation between a receveuse and a park goer.

In 1878, a gentleman living near Square Montholon wrote to the park service complaining about the receveuse who operated in the square. The "manners of the receveuse leave much to be desired," he wrote.[92] The woman had insulted him and his wife almost daily, he said, in "vulgar scenes" about which he had complained several times to the city administration, yet to no avail. Moreover, just the other day, in the company of his family, he wrote, he encountered her as he had before in a most "markedly disgusting" moment—relieving herself in the bushes. "Even the gardeners are complaining of it," he continued, "each day they find the proof of the fact and must remove it." If this were not enough, he suggested her intentions in the neighborhood were altogether suspect, adding "women of this sort should not have this type of employment because it is only a pretext to exercise their other Commerce [sic]."[93] The couple's contempt for the receveuse is graphic and palpable in the letter. Since the park service was concerned with the politeness of all workers in the park, it launched an inquiry into the incident. A park service inspector inquired about the woman with the brigadier guard who was responsible for the guard service in the area. "It is true," wrote the brigadier in his response to the inspector's request for more information, "that this woman *entered* into the bushes and *relieved* herself, but the guard *noticed* her immediately and *told* her she was forbidden to do so and she has since complied."[94] The brigadier continued writing that, "as for the use of vulgar language, she is often provoked by persons who refuse to pay for chairs. Although such refusals do not warrant an insult, this happened only occasionally and without ever giving way to any greater disturbance."[95] After having defended the receveuse, albeit in a measured way, the brigadier concluded somewhat flatly and without explanation that "this receveuse has not returned to the square in eight days."[96] It is unclear whether the receveuse ever returned to the square, or if any disciplinary action was taken against her by the park service. It is somewhat clearer, however, that the brigadier guard felt that the complaints were exaggerated, and that he, a fellow park worker, took her part in the matter revealing a certain esprit de corps among *all* those who depended on the squares for their livelihood and confronted similar unfair or condescending treatment by the residents of the area, particularly in the wealthier sections of Paris.

Similar conflicts and negotiations arose in the Square Louis XVI in the nearby Eighth Arrondissement between the receveuses in the square and some of the local society women. On 11 July 1885, Mme. Trouble, a woman of position and means living at 11 rue Vignon, sent a letter to the prefect of the Seine complaining of the "brutality" of Mme. Blin, the receveuse in the square.[97] On the same day, the commissioner of police for the quartier forwarded a policeman's daily report in which the officer noted that he witnessed the wife of the consul of Sweden and Norway having likewise been made "a victim of the nasty behavior of the same receveuse."[98] The park service quickly launched an inquiry into the incidents and, when it could confirm that the receveuse had indeed acted inappropriately, her concession was revoked. Just under a year later, complaints against Mme. Lacoste, a subsequent receveuse in the same square, came to the attention of the park service. A certain Doctor Hirtz had written to the *ingénieur en chef* of the park service to complain that his wife had reported to him that Mme. Lacoste had been "most improper" towards her on a recent visit to the square. In his report, the inspector for the park service indicated that Mme. Lacoste had only been working in the square for three weeks and during that time there had not been any complaints about her behavior. There were, however, witnesses to this particular altercation about which Doctor Hirtz had written. The witnesses apparently related some of the exchange between the two women because, when confronted with their testimony, Mme. Lacoste adamantly insisted that she had never said, "'So, you don't even have two *sous* in your pockets.'"[99] (Presumably the disagreement came about as a result of Mme. Hirtz's inability or refusal to pay for one of Mme. Lacoste's chairs.) In an effort to give satisfaction to Doctor Hirtz's complaint but also to treat Mme. Lacoste fairly, the park service determined to relocate Mme. Lacoste's concession to another nearby square and discipline her with a four-day suspension rather than rescind the agreement outright.[100]

Women who ran the small concession stands within the squares faced similar problems with the businesses in the surrounding neighborhood. One conflict involved a concession within the Place des Vosges, a nearby business, and much of the neighborhood. In 1884, M. Faucheux, an owner of a pastry stand on the corner of rue de Béarn and rue des Vosges (now rue du Pas de la Mule) that also offered ice cream to patrons, wrote to Alphand directly to complain that Mme. Felten's toy stand within the park was engaged in unfair competition. Faucheux claimed that Felten was illegally selling ice cream out of her stall. He asked Alphand to prohibit Mme. Felten from continuing to sell the treat which cut into his business located across the street. Alphand responded in a letter to M. Faucheux: "Mme. Felten was properly authorized to sell not only games, as you have supposed, but sweets, cakes, and the general kinds of refreshments that the *mères de famille* who frequent the park require for their children; you have therefore been misinformed about the nature of the authorization accorded Mme. Felten, who it seems to me has not surpassed the limits of that authorization by selling ice cream to children. I would add that the sale of these inexpensive refreshments would not be able to exert any substantial [negative] impact on the profits of your business, nor could it result in any serious material loss. In consequence, it is my estimation that there is no cause for action on your request."[101] Alphand's response, however, failed to stop M. Faucheux's campaign against Mme. Felten's operation. The following August, M. Felten contacted Alphand on behalf of his wife, saying that M. Faucheux, who calls himself a *"glacier"* (a professional ice cream maker) but in truth is just "an unfortunate like us who turns a *sabotière*" (a small ice churn and cooler) has lodged three complaints already and was preparing another. M. Felten asked Alphand to hear their pleas that this man might "leave us in tranquility in our square."[102] Even others in the neighborhood became involved in the drama. Several teachers from the Institution Galtière, École Communale, and Écoles des Frères (all schools in the area) joined the fray and sent a letter to the park service on behalf of Felten.[103]

Faucheux's complaints against Mme. Felten became well known within the administrative offices of the park service, although they repeatedly came to naught as the agency consistently defended the woman's interest. In response to a request made by Faucheux the following year to move Mme. Felten's kiosk so that it was not directly across from his business, the park service again took Felten's side. Faucheux had claimed now that Felten was selling exactly the same ice cream as he, and at the same price of five centimes a cup. The sous-inspecteur answering the request wrote in his report that moving her to another place in the park was simply out of the question; besides, the other concessionaires in the square, who might trade places with Mme. Felten, sold much the same product as she did. He concluded by saying, "In summary, we are of the opinion that Mme. Felten should remain in the position she occupies and sell ice creams at five centimes a piece as in the past."[104] In an addendum to the report approving the decision, the a senior engineer in the office overseeing concessions wrote, "As we have already commented in our preceding report dated 9 July 1884 in response to the first complaint of M. Faucheux, it seems to us that from the moment Mme. Felten was formally authorized to occupy a kiosk in the Square des Vosges [sic], to *sell sweets,* there was no reason to forbid her from *selling ice cream.*"[105] A year into M. Faucheux's battle against the concessionaire, the park service's support of Mme. Felten had not wavered, and officials appeared to be weary of Faucheux's repeated complaints.

Again in 1886, the park service received another letter of complaint against Mme. Felten. This time it was signed "M. Roche" and lacked any street address. The complainant wrote that "the guard in the square has authorized Mme. Felten to dig a sump hole beneath her kiosk, in which to throw water, salt and other ingredients used in her ice cream business."[106] The author's mention of the guard's actions suggests Mme. Felten may have had the support of her fellow workers within the square. "These waters have no drain," M. Roche wrote. Then, as if he were aware of the park service's abhorrence of damage to the trees and plants, he continued detailing the damage. The wastewater was full of saltpeter, he wrote, which "burns the soil and eats the roots of the trees. The elm tree to the right of the kiosk remains yellow year-round. Moreover, three other trees nearby appear to suffer as well in a way that might be occasioned by leaks in the underground gas pipes caused by these ingredients."[107] The inspector's report on how to handle the complaint noted that there had been an investigation of M. Roche's claims of damage in the square and that the park service concluded that "nothing that was said in the letter is true. No sump hole exists and the trees are in good condition." He closed his report writing, "One cannot help but see in this, the continuation of the war between the two ice cream makers, Felten and Faucheux. The letter appears to have no more value than an anonymous letter. There is, therefore, no consideration to be given in this matter."[108]

The experience of women concessionaires in the squares and parks of Paris is testament that the program of establishing fully public greenspace throughout the city, begun during the Second Empire and extended through the turn of the century, was not a socially painless process—certainly not at a time when class and gender distinctions remained rigid, or when women's engagement in commercial competition was resented and resisted. Wealthy Parisians, women in particular, often found themselves sharing their neighborhood greenspace with guards and workers whom they perceived to be their social inferiors. They could easily avoid sitting beside these "interlopers" to their quartier on a park bench, but if they wished to take a seat in the shade and no benches were available, they would be forced to deal one-on-one with a workingwoman within the park. These encounters were not always cordial or contained. Perhaps these women of means resented that they had to pay for a chair in what they considered their own backyard or that they were forced to engage a workingwoman who was not in their own service. Whatever the reason, they regularly and boldly challenged the licensed receveuses, and sometimes even refused to pay the most insignificant of fees—an inconsequential sum for these wealthy women, but likely essential to the household economy of the receveuse. The social tensions and confrontations that occurred in some of these greenspaces, and the necessary negotiations which inevitably involved the park service, were thus infused with class differences between park employees and residents of those more affluent areas. In tandem with these class tensions, or as part of them, ideas about women's "proper" domestic role in society, and the way in which the receveuse seemed to flout convention by working outside the home to earn money, no doubt simmered beneath the surface of many of the encounters.

Relations between male business owners outside the borders of the park, who resented competition from within the park, and concessionaires were equally contentious. If a female concessionaire's name came up in a complaint, her best hope lay in a reasonably fair system of review in which the park service attempted to gather independent information to confirm the veracity of complaints and help determine an appropriate response. Just as it might do with guards who had become overly friendly with the neighborhood residents and frequented taverns more than the square, the officials of the park service sometimes chose to relocate a receveuse rather than void her concession based solely on a supposed confrontation a park goer. Moreover, they might even defend the interests of a businesswomen operating legally within the greenspace against businessmen in the quartier who sought to shut down her enterprise. The park service's evenhanded stewardship of the greenspaces, rooted in practicality and function, intersected and advanced a sense of the democratic nature of the squares of Paris (democratic, but limited to those who abided by city statutes). If the spaces became socially homogenous or delineated, it was not owing to any official policies or practices on the part of the park service.

Those for whom urban public greenspace was also a workplace (guards, cantonniers, and concessionaires) derived considerable benefit from their employment in those spaces. Former military men had a viable option for a working retirement. They could feel secure in knowing, as well, that if they could not work or worse, died, their wives and families might yet have some means of support either by obtaining a small concession or through assistance from a well-funded mutual aid society. Some enjoyed comfortable housing within the park space, an attractive bucolic alternative to the cramped urban housing their income might afford them elsewhere in the city. Women, too, who could run their own enterprises and turn a small profit were assured of a degree of support from the Service des Promenades et Plantations. For entrepreneurs with more ready capital, the larger concessions in the park provided a fairly safe and lucrative investment. Work in the parks also often connected one to a community of fellow park employees and operators, and, in the case of the smaller squares, to a community of one's neighbors, sometimes in a position of high regard in the quartier. In short, for those who made their living in the parks and squares of Paris, the work was good, not onerous, and the income was steady and reliable.

Still, employment in the parks was not without its significant challenges. The park service, although a generally fair player, exercised considerable control over their employees' work lives and businesses. No concessionaire, large or small, enjoyed complete autonomy, and the guards and cantonniers were subject to rigorous, sometimes withering, scrutiny of their job performance, both by the park service and by a rather expectant and demanding public. Working in public space meant one had to serve many masters. An easy rapport with those in the neighborhood was not a given, nor was respect for positions that represented the municipal government. Women workers faced intense resentment and even mistreatment by park visitors and neighborhood businesses. Indeed, the social landscape of these greenspaces (particularly the smaller squares) could be extraordinarily difficult to navigate given contemporaneous understandings of class differences and gender roles. For these workers, the employment was good; the workplace was a constant challenge.

The men and women who depended on Parisian greenspaces for their livelihoods contributed immensely to the democratic and popular nature of the parks. They were a bridge between city planners, reformers, and engineers and the city population. On the one hand, it was their job to implement the program of the development of fully public urban greenspace. They were to insure that all might use the park without damaging it and without preventing the enjoyment of others. The concessionaires were to provide services relevant to the desires of the visitors to any given park or square—toys and sweets for the mères de famille and children who frequented the space, or entertainments and cuisine for the others. On the other hand, the very presence of these workers helped democratize the spaces, particularly when working-class park employees inhabited greenspace along with wealthier city residents. As social navigators and arbitrators, they, along with the public and the Service des Promenades et Plantations helped work out the meaning of these spaces, and the role they would play in city life. Men and women who made a workplace of the parks and squares of Paris contributed, through their spatial practices (how they moved about the parks and squares in the performance of their jobs and how they were received by employer and park goer alike) to the creation and understanding of fully public urban greenspace, yet they were not alone. Park goers, the intended users of these engineered greenspaces, contributed in equal measure, if not more, through their own spatial practices and communal formulations, which were centered on Parisian greenspace.

JUST AROUND THE CORNER
THE NEIGHBORHOOD AND THE URBAN PARK

••••   ••••

How is it that under a republican government, and particularly under the governance of a republican Municipal Council, the residents of the neighborhood of the Place royale [Place des Vosges] who, during all of the Empire and the presidency of M. Thiers, compared to other neighborhoods had had the pleasure of hearing on Sundays and Thursdays, in the Place royale, a concert given by the troops, but are now deprived of this pleasure while it is preserved in aristocratic neighborhoods? And how is it, that the garden, ornamented with four fountains and jets of water, has been deprived for so long . . . of the delightful spectacle of spouting water in its basins?

Here, sir, are the facts that I believe should be brought to your attention and that seem to me to be the result of ill-will on the part of some authority.

In the hopes that it will render us justice, I count sir, on your official solicitude with regard to that which concerns our vested interests [in the square], grandfathered rights, which we hold very dear indeed.

—LETTER FROM A. PICHOT, SCHOOLTEACHER TO AN
UNNAMED MUNICIPAL COUNCILOR, 1877

••••   ••••

Workers in public parks and squares during the second half of the nineteenth century interacted daily with local residents and others who used greenspaces. As much as those workers may have considered the park or square to be their workplace, for Parisians who frequented the spaces it was much more. For them, greenspace came to be an extension of their living space—the home, and a locus of community activity and formation. In sharing use of local greenspace, fellow city residents helped to define common and competing interests among themselves, which often led to the formation of smaller communities within neighborhoods. In this sense, the parks and squares triggered a kind of spontaneous and direct social formation that anthropologist Victor Turner referred to "existential *communitas.*" "Rituals of *communitas,*" according to Turner, both express and reinforce a sense of community when regularly repeated, revivifying the feeling of unity and defining the participants as a community unto themselves.[1] Turner used the term "communitas" rather than "community" to distinguish it as a "modality of social relationship" as opposed to an "area of common living"—a way of communing through experience rather than a location.[2] He argued that the spontaneity of this phenomenon cannot be sustained for long since it quickly gives way to an organizing impulse, or what he referred to as "normative *communitas.*"[3]

The ostensibly natural spaces of the parks and squares, liberated from the rigidity of the physical forms of the built environment, and challenging, as they did, social expectations of city life, were fertile ground for the development of Turner's spontaneous, organic process of communitas. That kind of "communing through experience" occurred as often within the public park as it did in the areas adjacent to it. Residents living in and around greenspaces and park visitors sharing the space came together (or clashed) over issues of access, use, management, and design, issues brought about by the introduction of greenspace in closer proximity to their homes. With large and small parks and squares scattered throughout the city, it was easy for neighborhood residents and communities to feel a personal connection to these spaces of communitas. These were locations where many individuals often spent part of their day, interacting with their neighbors. This personal connection engendered a proprietary sensibility among residents that was expressed in communications (often collective) with the *Service des Promenades et Plantations.* This communication increased over time, and neighborhood residents became more organized and more specific in their demands. Moreover, the general responsiveness of the park service to these wide-ranging communications, and the inherently flexible nature of greenspace (a gate could be opened, a bandstand built, plants watered, a new path constructed) may have furthered this collective sense of ownership over the public spaces, distinguishing them from other kinds of urban spaces, such as private commercial and residential properties.

To comprehend this dynamic environment, it is helpful to sort out who the park goers were, who moved within these habitats, and what were their interests and concerns. Having considered those people who worked in the parks and constituted the public face of the park service, I will now focus on those who derived no direct economic gain from the spaces: park visitors. As there are lacunae in the historical record, my approach here is nonlinear, and not strictly chronological, yet in many ways the most effective means of rendering an image of the human landscape of these spaces. Taking a lead from Michel Chevreul and Georges Seurat (the former's laws of contrast in color affected the art of the latter), we can follow something of a "pointillist" approach. Moments and places in time, dots that at first seem incongruous and insignificant, in composition paint a vivid portrait of life and communitas in the public greenspaces of nineteenth-century Paris.[4]

Of all of the groups that frequented and influenced the public greenspaces of Paris, *riverains* and *habitants* comprised the largest and most diverse. Riverains were those whose property or business abutted the greenspace, and habitants, all of those living in the surrounding quarter. These park users sought to personalize and make relevant in their lives the greenspaces they now found closer to their homes. They wrote letters, petitioned, and eventually became quite organized, all in an effort to make known their wants and needs regarding public space, and to shape the local parks or squares. In the eyes of the park service, those who resided or worked around the park, whether adjacent to it or nearby, constituted the single group with the greatest personal interest in the greenspace, owing to their close proximity. This perception however, did not always translate into carte blanche for those park users. The park service sought to accommodate *all* of the park visitors' interests while maintaining the integrity of the greenspace and, as a consequence, often rejected some requests regarding access, use, management, or design of the space.

Questions concerning access and park gates arose in 1865 during the construction of the Square Louis XVI (4,183 square meters), which, once completed, would enclose the *Chapelle Expiatoire* on the new Boulevard Haussmann. Louis XVIII had ordered construction of a chapel there during the Restoration as a demonstration of national atonement for the executions of Louis XVI and Marie Antoinette during the Revolution. At the time, the surrounding property belonged to the state and was administered as such. During the Second Empire, the national government worked in partnership with the municipal government to manage these kinds of state-owned properties within Paris. On learning of the proposed enclosure, landlords and renters whose homes stood on the eastern side of the square joined forces. In a letter to the administration, they called for a gate that would allow them to cross the park on the way to the Boulevard Haussmann. They had been able to move freely around the chapel on their way to rue Pasquier, but now, the square interrupted that path and caused them either to go around it via the rue des Mathurins, or enter it at the corner on the same street. (See map 1.) In their petition, they cited *"préjudice"* and *"l'expropriation indirecte,"* legal terms that hinted at a desire to ready their case for one of the increasingly generous expropriation juries whose large awards were well known.[5]

The chief architect of the park service, Gabriel Davioud, handled the petition since his office was overseeing the work on the square, given the need to preserve the integrity of the architectural structure at its center. The design of the square, argued Davioud, was carefully conceived to account for the odd position of the chapel on the bias within the space. He added that the city tried always to avoid treating a square as if it were the kind of public passage that the petitioners wanted by routinely constructing circuitous paths to extend walking distances. The rather obvious legalese in the petition seemed to surprise Davioud most: "If they wish to speak of material loss and indirect expropriation, they would first have to establish *the right* [legal] they hold to traverse the garden of the Chapelle Expiatoire. But this right does not exist in any form and the state has never allowed the public to traverse soil that it owns except by way of simple tolerance."[6] According to Davioud's logic, the city had not expropriated anything because nothing had ever been legally owned. The concept of a *right* to air and light, which Adolphe Alphand and others had held as a moral right, had become so enmeshed in the promotion and rhetoric surrounding the projects that the residents of this neighborhood now took it for a legal right, much to the consternation of administrators such as Davioud. The public's perception of "rights" to city space informed many interactions with the Service des Promenades et Plantations regardless of the neighborhood, and persisted well into the decades of the Third Republic.

Across town, in the working-class Nineteenth Arrondissement, the newly constructed Parc des Buttes Chaumont (231,567 square meters) led to similar communications between habitants and city officials. In 1866, when construction on the park was underway, residents and business owners on the rue

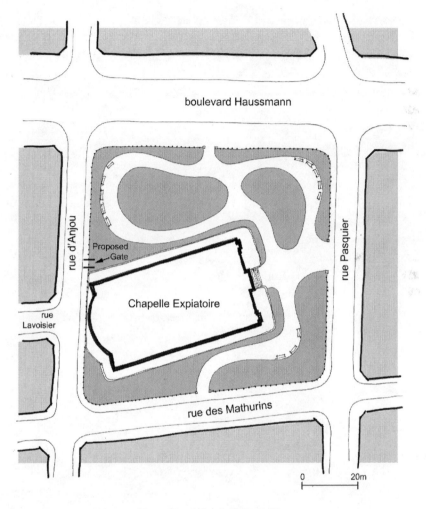

Map 1. Square Louis XVI, ca. 1865.

des Alouettes sent a petition to the prefect asking for the creation of a new entrance to the park at the end of their street. They had lodged a complaint months earlier expressing concern that the park would slow traffic in the neighborhood and harm their businesses. Now, they requested a new entrance as "compensation." After reviewing the matter, the park service produced a report in which it contended that, before the construction of the park, rue des Alouettes fed into the rue de Chaumont, an unimproved and practically impassable road. It connected to the larger rue de Crimée only by way of a narrow cart path descriptively known as the *Chemin du trou d'Enfer,* or "Hell Hole Road."[7] With the installation of the new park, rue des Alouettes now connected with the improved peripheral park road (rue Botzaris) which in turn led directly to the rue de Crimée on one end and the rue de Puebla (rue Pyrénées) on the other end. (See map 2, A.) This was a much better route for transporting goods by wagon. As for easy access into the park, the report acknowledged that it would be ideal if every street that abutted the park could have its own entrance. However, as this was not the case with other public gardens such as the Tuileries, located in an area with an "enormous" population, residents could not expect it in a "nearly deserted" area like Buttes Chaumont.

According to the park service, there was an entrance to the park approximately 150 meters from the end of rue des Alouettes, so the petitioners found themselves in a relatively advantageous position should they wish to enter the park.[8] Although the report put an end to hope of a transverse road through the park and a carriage entrance at this location, it was not the last effort the residents of rue des Alouettes mounted to increase accessibility to the park

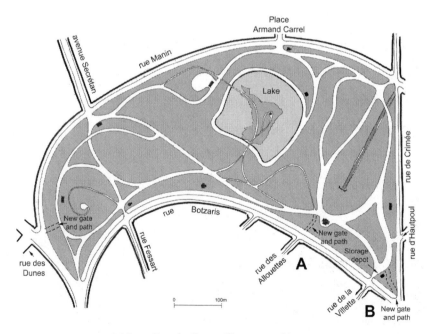

Map 2. Parc des Buttes Chaumont, 1867–93.

space. Alphand's 1867 rendering of the Parc des Buttes Chaumont shows that, subsequent to the 1866 petition, the park service opened a gate with a footpath at the very point where the rue des Alouettes met the rue Botzaris, precisely where the residents had requested.[9] A later report, produced in 1879, concerning another local group's request for access to the park at a different location, explained how the 1867 rue des Alouettes entrance had come to be. According to the report, three additional *guichets,* or pedestrian gates with pathways, had been opened to the park since its initial design, and all three of the added gates had been opened "following the request, verbal or written, of the local residents and at the directive of M. le Directeur des travaux de Paris."[10] Despite a mix of successes and failures, the residents in the neighborhoods surrounding the park persisted in their efforts to gain access to what they reportedly had begun to refer to as "their Bois."[11]

Those Parisians who resided on rue des Alouettes were not alone among the habitants around the Parc des Buttes Chaumont who collectively petitioned the administration during the construction of the park. Although their street was one block closer to the rue de Crimée, the residents of rue de la Villette likewise petitioned the park service for a carriage road through the park, and a similar report was generated to explore the issue. The residents argued that the narrow, far-west corner of the park interrupted one of the roads connecting the former villages of Belleville and La Villette.[12] (See map 2, B.) This, they claimed, jeopardized communication of goods and people between the two locales. In its report, the park service contended that the construction of the park had subsumed only ninety meters of the rue de la Villette, and that the new layout of the paved roads presented a much better scenario for circulation. Although it may have been a straight line, that section of the road was uninhabited, poorly maintained, and had a steep 10 percent grade. The residents of rue de la Villette needed only turn right at the end of their street and proceed a few meters to reach rue de Crimée. This added only about fifty-five meters to the length of the trip and was much easier on horses and pedestrians as it reduced the grade to between 4 and 5 percent.[13] Moreover, as was not the case for the residents of the rue des Alouettes (in 1866 at least), there was an entrance to the park right at the end of rue de la Villette. In both cases, the park service rejected the idea of making significant alterations to the park design to accommodate street traffic, particularly when the rationale was dubious, yet it placed real value on ready access to neighborhood residents entering on foot.

Petition campaigns continued as Alphand's team hurried to complete construction of the park, with the opening scheduled to coincide with the *Exposition Universelle* in 1867. Some neighborhood groups were quite vigilant about developments in and around the park. In early 1867, the residents of the rue des Alouettes united with those living on the rue du Plateau to complain about the pace of the work on the park and surrounding roads. This time, they sent their petition directly to Napoleon III, a potentially effective strategy given the emperor's recent, well-publicized personal intervention in the fight over the Luxembourg Garden, and his particular interest in the Buttes Chaumont project.[14] Reference to the petition appeared in a report submitted by a lead engineer on the project, Jean Darcel. Darcel was responsible for managing the monumental engineering project of building a public park on one of the most polluted and degraded sites in Paris, while at the same time responding to the complaints and concerns of residents in the area. Although the precise demands of the petitioners and the emperor's response are unknown, the petition's existence sent ripples through ranks of the park service and angered Darcel. In his report on the matter (a response to an internal memo), the engineer did not mince words about this latest complaint (embarrassing perhaps as it had involved the emperor) and his frustration with the neighborhood residents. His response reflects the scrutiny local residents placed on the project and their liberal use of petitioning to communicate their concerns. "The petition addressed to the Emperor," he wrote, "by the residents of the rue des Allouettes [sic] (et du Plateau) is new proof of the absurdity of the populations of the communes joined to Paris who believed that the law of annexation made the municipal administration into a fairy, able to transform crooked, muddy, rutted, unpaved lanes with a wave of the wand, into Boulevards des Italiens."[15] Indeed, the lack of good roads in the area slowed the work on the park. And these working-class residents did not stand by watching passively as their neighborhood was transformed; rather, they asserted themselves and brought pressure to bear on the city agency and engineers executing that change, something that often surprised and rankled those officials.

This kind of local community interest in neighborhood parks and squares and engagement with municipal officials continued into the 1870s. The Service des Promenades et Plantations, under Adolphe Alphand's direction, often had to sort out what kinds of requests, in its view, could and should be accommodated and which did not justify costly or permanent alterations to the space. It attempted also to determine who among the petitioners had what it deemed a vested interest in the greenspace. In 1875, a small group of Parisians who lived near the Square Parmentier (now Square Maurice Gardette) in the Eleventh Arrondissement petitioned for an entrance on the rue Guilhem on the eastern side of the square. The square had three existing entrances: two on the corners of rue Guilhem and rues Lacharrière and Rochebrune and one at the opening to rue Renault. (See map 3.) The park service rejected the request based on the proposed gate being able to serve only the approximately eight to ten households (signatories to the petition) and those residents needing to walk only forty meters in either direction to arrive at one of the existing corner gates.

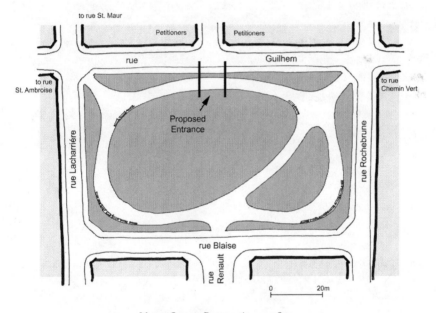

Map 3. Square Parmentier, ca. 1875.

Undeterred, the habitants petitioned again the following year for the same new entrance. This time, they had curiously increased their number to fifty-five, and they added reasons of public safety to their argument, claiming that the presence of a carriage manufacturer on the nearby rue Chemin Vert made entering at that southeast corner of the park dangerous and thus justified the new gate. Again, the city declined the request. The park service official answering the petition carefully mapped the addresses of the petitioners and found that only fourteen of the fifty-five actually lived on the two small passages that the new gate would serve. The remainder lived farther away on rue Saint Maur between rue Saint Ambroise and rue Oberkampf. As far as the park service was concerned, these additional habitants had no direct interest in the new gate because even with it they would naturally continue to enter the square through the nearest gate at the corner of rue Lacharrière.[16] Significantly, the original group of petitioners had enlisted the help of others in the quartier. Those additional petitioners had little reason to care if there was a new gate since it would not affect their ability to access the space. Yet, in joining the original petitioners, they acknowledged shared neighborhood interest in the space and the right of all residents to easy access. Although this petition was not successful, the square had been a catalyst for increasing neighborhood unity. That same sense of shared interest and unified approach grew into larger, more organized, neighborhood-wide efforts, which became more and more common in Paris in the later decades of the century.

In the 1880s this kind of organization, coupled with changes in population, transportation, and work schedules in the city, resulted in the park service seeking new ways to reconcile their desire for dedicated greenspace with the realities of modern life and the wish to accommodate residents' practical need to move through the park on the way to and from work. Naturally, this was not as critical in the case of small squares such as the Square Louis XVI, where the detour was only a few meters, but when it came to larger parks such as Buttes Chaumont (231,567 square meters) or the Parc Montsouris (154,640 square meters), it was a concern and a challenge. Indeed, in some quartiers, pedestrian access to the greenspace became central to workers' ability to get to and from the workplace. Prior to the installation of the Parc des Buttes Chaumont, the area constituted a formidable obstacle for workers who lived on one side of the space and worked on the other. The gangs who inhabited the no man's land of the abandoned quarries and garbage dumps made crossing parts of the area a risky and dangerous proposition, and circumventing the buttes would add significantly to workers' transit time. With the inauguration of the park, however, workers in Belleville and la Villette had a new, safer way to get to the workplace. The park service acknowledged this particular use of the space in an 1882 parks administration report, even as it attempted to balance concerns for public safety. In it, engineer Seilheimer responded to a request by a municipal counselor to extend the park hours to midnight all year long. (At that time, the park closed at 8 p.m. from December to January; 9 p.m. in February and November; 10 p.m. in March, April, and October; and 11 p.m. from May to October.)[17] This request was actually a reformulation of a proposition from the previous year that would have kept the park open all night had it been approved. Seilheimer asserted that the park service endeavored as much as possible to set park hours specifically to allow workers to traverse the park easily at the start and end of their long workday. However, he insisted that in a park like Buttes Chaumont, which lacked sufficient lighting, extended hours would simply be too dangerous and would potentially expose workers to thieves and even murderers who might be lying in wait at night.[18]

In the same year, the Parc Montsouris in the Fourteenth Arrondissement presented a similar problem involving space, safety, and Parisians returning home from work. The city purchased the majority of the land on which the park was constructed from two rail-line companies: the *Chemin de Fer de Paris à Limours* and the *Chemin de Fer de Ceinture*. Easements were included in the sale, which allowed the former to maintain its Sceaux station within the boundaries of the park, and the latter to pass perpendicularly beneath the first with its Glacière station just outside the eastern side of the park. Commuters, who disembarked from the Sceaux train and wished to change to the Chemin de Fer de Ceinture, and vice-versa, routinely took a shortcut through the park. (See map 4.) However, after closing hours, only the Sceaux gate remained open. Unlike the case of the Square Louis XVI, nearly twenty years earlier in which the park service determined that the square was not a public way, in the larger Park Montsouris the agency attempted to accommodate the transit of these passengers. The park service agreed to allow both the Sceaux gate and the Glacière gate to remain open year-round until 11 p.m. (summer hours) and would install street lamps along the path between the two gates. At the end of their evening shift, the guards were instructed to set up barriers all along the path and post signs pointing the way for commuters.[19]

For the city, it seemed like a good solution. However, problems soon arose as more and more passengers who had worked late attempted to take a different shortcut across the full length of the park in the direction of Montrouge and found the gates on that side were not opened. According to an 1882 park service report, on reaching the gate, passengers responded with a great deal of anger, "shouts, threats and eventually climbing over the fence."[20] The disturbance

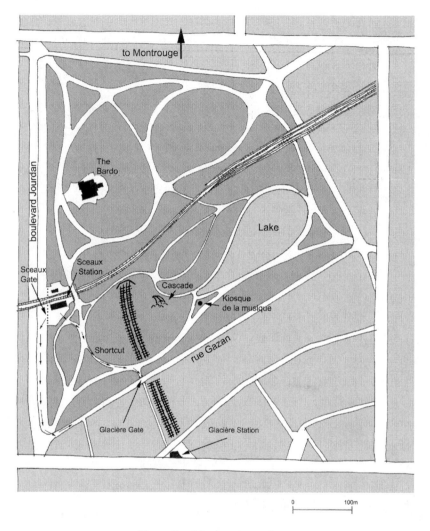

Map 4. Parc Montsouris, ca. 1875.

was such that the park service clearly had to alter its plan. It could certainly order that the entire park should remain open all year long until 11 p.m., but that would mean extending the shift for the guards and increased costs. That might not have been a problem in the summertime when the added expense could be justified because more of the local public was in the park anyway, but in the winter, the considerable hardship on the guards, particularly on cold winter nights, seemed too much. Moreover, although many of the squares in the city were in the process of being equipped with lighting, Montsouris had only recently installed lighting along the path between the two rail stations. This would have meant that scores of people—some simply lost and some with malicious intent—would have been milling about the park in the darkness of night. Public safety could be at risk in such a scenario. The park service backed away from its early plan, abandoned the shortcut, and created an exit outside of the park limits for the Sceaux station. Passengers were to follow the sidewalk along Boulevard Jourdan to rue Gazan to make their connections. (See map 4.) This was a better situation generally, according to the park service, since the distance was only 140 meters, and the sidewalks along the Boulevard Jourdan were fairly level and covered in asphalt, while the park paths were steep, slick, and particularly dangerous in snowy and icy conditions. The park service attempted to accommodate the need travelers had for ease of access but found that, in this case, questions of public safety prevented them from doing just that. As the population of Paris grew, so did the number of people wanting to use public parks in varied ways. Increasingly, rights of access often gave way to questions of public safety, particularly with the introduction of automobile traffic in the largest parks.

Toward the turn of the century, well-organized petitioners sometimes raised seed money to eliminate the potential for a denial based on budgetary constraints and worked through their local municipal counselors. In 1890, a group of neighborhood residents who lived next to the Parc des Buttes Chaumont between the Place Armand Carrel and the rue de Crimée deposited 580 francs in the municipal treasury to pay for the construction of a new guichet located west of the place on rue Manin. The guichet, which would be designed and built by the city contractors, was approved.[21] Three years later, municipal counselor Armand Grébauval proposed opening another guichet on rue Manin near the corner of rue Bolívar. Grébauval claimed that the gate would "certainly be useful and would very much facilitate access to the park for people living in the neighboring streets."[22] This gate would open up the park to the residents in the crowded, working-class housing located along the rue des Dunes and the Impasse du Puits. The park service approved and built the gate. (See map 2.)

Again in 1893, Grébauval and fellow counselor Charles Bos joined forces on behalf of the *bellevillois* (local Belleville residents) to push for yet another gate at the corner of rue Botzaris and rue de Crimée. This entrance, however, required a great deal more consideration, according to the park service, since the proposed path would run straight through a storage depot and tool shed used by park employees. The heavily wooded area in question also contained a number of mature trees that would have to be moved or cut down, if the entrance were to be approved. Engineers proposed two solutions: move the depot a few meters to the side, or leave it in place and run the public path around it.[23] Officials of the park service were generally cool to both ideas. In an 1896 report, the chief landscape architect assigned to the project wrote that the new gate was "not of great utility; it doesn't appear to have any other *raison d'être* except to give access to persons in Belleville who, coming from the rue Crimée, don't like taking the Botzaris sidewalk to the gate de la Villette."[24] The distance from the corner to the de la Villette entrance was a mere forty meters, and there was another guichet not far away on the rue de Crimée at the intersection of rue d'Hautpoul. Nevertheless, the park service approved and built the new gate with a path going around the depot. (See map 2, B.) Organized residents working through their representatives successfully capitalized on the idea that greenspace in the city existed for their benefit and use. They were sometimes successful in obtaining multiple entry points and greater access to the interiors of public greenspaces in this way, particularly in the larger parks.

Inasmuch as the residents pressed for increased access to the greenspace in their neighborhoods, some groups also petitioned to use the space to hold community events, often fundraisers, or other social events. These gatherings could function much like Turner's "rituals of *communitas*," yet the park service remained concerned that these events would not impinge on the rights of another group to use the space. Very often, the location and size of the greenspace, and the nature of the event had more to do with obtaining approval than which group or community had made the request, although that was certainly given some consideration as well. In 1887, the mayor of the Fourth Arrondissement forwarded a petition, signed by a group of carnival merchants requesting permission to hold a *fête foraine* (carnival) in the Place des Vosges (12,377 square meters) to benefit the arrondissement's school fund. Alphand's office quickly rejected this request, in spite of the worthy cause. In his response to the mayor, a park service representative cited the inconvenience the carnival would cause to the neighborhood residents and the public health danger of holding an event in such a relatively small public greenspace: "In establishing, at great cost, squares in diverse *quartiers,* the city of Paris wanted to create places of retreat where the population could find calm and tranquility along with verdure. It would not be wise of the Administration [Service des Promenades et Plantations] to allow them to be invaded by boisterous festivals."[25] Furthermore, he continued, a recent commission (of which Alphand was a member) produced a report at the request of the *Conseil d'hygiène,* which called attention to the health dangers of fêtes foraines in the city. The report indicated that the deleterious effects of such events were even more grave in close, confined spaces, precisely like the Place des Vosges, and its findings resulted in a ban on all such festivals within city limits.[26] The park service had certainly not been categorically opposed to fêtes foraines in the past, provided the space was large enough. Three years earlier, in 1884, a workers' mutual aid society, the *Société laïque d'appui fraternel,* composed of local merchants from the area around Buttes Chaumont, had petitioned the city administration successfully for permission to hold a fête foraine in the park. The carnival, they said, would benefit the "honest" local workers who had suffered during the recent economic "crisis" that had so affected the population of the Nineteenth Arrondissement.[27] In this way, the park became a very real center of community formation and communitas for both the residents and families in the immediate neighborhoods, and for the larger community of workers in the area.

In the last decades of the century, musical performances in public greenspaces developed rapidly as another medium of communitas. They had occurred sporadically in the 1860s but became increasing popular and grew into something of an institution by the 1910s. In nearly every square in every quarter of the city, from the posh west side to the working-class arrondissements in the northeast, a *kiosque de musique* stood in the midst of neighborhood greenspace. Many a square had its own local orchestral society made up of neighborhood fathers, sons, brothers, and uncles. These musical events brought the residents of the quartier together several times a week to enjoy music played in the open air, often by their friends and relations, further cementing a sense of community. The park service regularly approved requests to hold concerts provided the organizers understood that the permission would not be automatically renewed, that it could be revoked at any time, that the organizers would be responsible for the expense of extra park guards if the size of the crowds required it, and that the concerts must be open to all and free—per Alphand's orders.[28] If there were multiple requests, the park service directed the different groups to work out a schedule agreeable to both.[29] By the 1900s, this kind of cooperation became even more critical as musical groups often toured the city performing on the bandstands of other quartiers. In 1907, the leaders of six separate orchestras sent a joint request to the city to hold summer concerts in the Place des Vosges. The groups came from the Third, Fourth, Eleventh, and Twelfth central and eastern arrondissements and presented a completed summer-long schedule with their petition to the city. Not only had groups become more organized in their interactions with the city, they also began to coalesce into a citywide community of Parisian musical societies.

Military bands and music were particularly popular in Paris at the turn of the century. The rousing and patriotic tunes fit the nationalist and bellicose mood of the time, and the anti-militarism of the Socialist Left leading up to World War I had yet to gather real steam. Kiosks throughout Paris welcomed the music, but the crowds often caused a headache for the guards charged with protecting the plants, and of course the hard-won grass. At a summer evening concert in 1901 in the Square Parmentier, the two guards on duty estimated the crowd to be nearly six thousand, and the damage was terrible. "[A] hazelnut tree was uprooted," wrote the Garde Géraud, "bushes planted just this year by the kiosk had their leaves pulled off, and the lawns were invaded by a considerable number of persons, some sitting, some standing, some lying down, and there were others still who were dancing to the music."[30] Similar scenes occurred in other parks as well, such as at the concert given by the *Garde Républicaine* that created a sensation in the Parc Montsouris on a pleasant day late in May of 1903. The guard report from the day estimated the crowd at 20,000 and said people seemed to pour in from every entrance and made their way to the lower part of the park where the kiosque de musique was located. The ten elderly guards, brought in for overtime duty that day, were powerless against such a surge of people. Just as the orchestra of the Garde Républicaine began its second piece, one park guard reported that a "torrent of humanity spilled onto the lawns, first around the area of the kiosk, then everywhere else besides, trampling the grass; pent-up on one side [the crowd] flowed back upon itself on the other in an irresistible backwash."[31] A dramatic telling of events, to be sure; nevertheless, the crowd that day was exceptionally large owing to the pleasant weather, and the guards were completely unprepared to manage it. They reported that the audience was larger and even denser than the crowds they often saw attending the fireworks on 14 July. "[T]o keep order at those events," they told their superiors, "there is a battalion of troops and 100 policemen." The park service should organize a similar force should "the music in question make itself heard in the park again" to prevent damage to the park and disorder or injury.[32] The musical performances in the parks had grown so much in popularity that there was a real risk of serious damage (beyond simply trampled grass) and again, a danger to public safety.

Parisians were as concerned with the maintenance and management of their greenspaces as they were with obtaining access to them and using them to serve the community. They monitored closely the condition of the parks and squares, and communicated their concerns to city authorities. In 1872, residents living near the Luxembourg Garden perceived a lack of sufficient care for their local greenspace. They chose as their spokesman M. Marceaux, a neighborhood resident, who was a retired *chef de bataillon* and member of the *Légion d'honneur*. Marceaux wrote a letter to the prefect complaining that the Luxembourg Garden sorely needed watering. He contended that "[t]he absence of watering during hot weather when the garden is invaded with a large number of the public causes a great quantity of dust to be lifted into the air which is a considerable inconvenience to the park visitors and the children who gather there to engage in their usual play."[33] Marceaux also communicated that, "with regard to the social position and the fortune of the greater part of persons called to frequent the garden habitually, they constitute, more so than most in Paris, a population for whom well-being and the pleasures of a vacation are forbidden."[34] The park service sought permission from the *Sénat* (who still maintained jurisdiction over the part of the garden nearest to the palace) to install irrigation plumbing throughout the garden with twenty-two valves made to receive watering hoses like those in the Bois de Boulogne. The cost of installation was nine thousand francs, but the report indicated that the park service would add the yearly operating and personnel costs to its budget.[35] Marceaux's success in this matter had more to do with the public good inherent in the request than with his military background or status, since other petitioners of equally respected position were routinely denied if the benefit to the public was not sufficiently established.[36]

Concessionaires and landscape crews in the parks were another focus of complaints by local residents about the management of the parks and squares. One incident involved a *receveuse* in the Parc des Buttes Chaumont in 1896. In the summer of 1896, the municipal council received a letter of complaint from a M. Buchet who lived at 28 avenue de Laumière near the Place Armand Carrel in the Nineteenth Arrondissement. M. Buchet had written to his municipal counselor, Armand Grébauval, concerning the activities of a particular concessionaire as well as some guards and maintenance men—behavior he found quite objectionable. M. Buchet claimed that the receveuse near the bandstand by the lake prevented people from placing their own chairs in the shade beneath some very large trees, insisting that those spots were for people who paid ten centimes. This often sparked angry words between the park visitors and the concessionaire. In addition, the landscapers who worked in that area did their watering between 2:30 p.m. and 3:00 p.m. and in such a way that it created a veritable muddy swamp in and around the path, and the benches were too wet to use. Finally, Buchet complained that those same workers "are given to using the expression f——le camp [taken to mean *fous le camp,* or "get the hell out of here"] so as to upset the park visitors."[37]

Grébauval sent the letter to the park service, and Inspector Lion led the inquiry. Lion stated in his response that there had never been a report of park guards having to intervene in a dispute between this receveuse and the park goers. M. Buchet, Lion surmised, must be referring to an incident that had occurred in May when a different receveuse, Mme. Véron, and a park restaurant worker, M. Boisschier, got into a heated argument. Boisschier was in some way "bothering" Mme. Véron while on his lunch break. The argument was loud, and many insults exchanged. Exasperated, Mme. Véron asked a nearby park guard to conduct them both to the local police station so that she could file a complaint against Boisschier. Following the incident, an investigation by the park service revealed that Boisschier had been the source of many similar altercations in the park in the past; it ordered his dismissal from his position at the park restaurant.[38] The landscaping crew in the area was interviewed during the inquiry. They claimed that they were finished watering in that section by 2:15 p.m. every day and that they intentionally sprayed down the benches in the sun to clean them off. Those seats are dry within fifteen minutes, they said. As for the rudeness of the workers, Lion stated, "*Cantonniers* are ordered to always be polite to the public and they generally are, particularly Bourillon [a foreman] who is in charge of this part of the park. We have never received a single complaint about him."[39] Lion recommended a response letter to M. Buchet, but no further action in the matter.

Some years later, in the southern part of the city, a habitant in the neighborhood of the Parc Montsouris, M. Baronnier, wrote to the prefect of the Seine to complain about two issues he found very troubling. First, the path which he followed each day to catch his train at the Glacière station had had a barrier across it for nearly a month. He and others like him were forced to take a detour along a very steep "*casse-cou*" of a cement path in the park. Furthermore, the landscape workmen, wishing to cause mischief, often sprayed passersby and then denied having done so, causing great embarrassment to the park users.[40] The park service investigated the complaint and produced a three-page report complete with a map showing the path that had been closed temporarily to allow for necessary repairs, as well as routes it suggested M. Baronnier might take to make his train. As for the behavior of the workers, not surprisingly, the men denied having sprayed any park visitors; however, this kind of practical joke seems in keeping with other reports about these often irreverent and hard-to-control park employees. The park service official said he wished to pursue the matter further but found that he could not since it would require an interview with M. Baronnier but he had not provided an address so any follow-up was impossible.[41]

Whether Mme. Véron bullied non-paying customers, or Boisschier harassed workingwomen in the park, or the cantonniers reveled in irritating park goers and using vulgar language, or even whether M. Buchet or M. Baronnier were concerned citizens or difficult malcontents is not really as significant as what the reports reveal. These accounts indicate a lively atmosphere full of any number of scenes of human interaction in the public greenspaces—some amusing, some less so—among various elements of the neighborhood, and between social groups and classes who inhabited the parks. Buchet's and Baronnier's letters of complaint, and the related inquiry and interviews they generated, demonstrate, on the one hand, the lengths to which the park service would go to follow up on complaints it received from park users about what some might consider the more banal aspects of the operations of a park or square. On the other hand, the letters underscore how seriously some habitants viewed the quotidian in what they considered *their* parks.

Groups and individuals were equally keen on influencing the actual layout and design of their greenspaces. City residents from all parts of the capital tried in various ways to reshape the interior spaces of local public greenspace to suit their individual or collective needs. In some instances, the alterations were remarkable. Alphand received a petition in 1873 from a number of mothers who frequented the Square Montholon, bringing their children there to play. The women explained that they wished the park service to remove completely the cascade and pond that had stood in the middle of the square for nearly ten years because the water feature was a "subject of fear" for the health and safety of their children.[42] (See map 5.) Alphand ordered his engineers to devise a way to remove the water feature to "give satisfaction to the complaint of the *mères de famille*."[43] The team calculated that the cost of the removal would be significant at seven thousand francs. This cost was communicated to the prefect of the Seine, who approved the expense and, in a letter to the Municipal Council dated 4 December 1873, urged the council to work this expense into the coming year's budget as "Improvements to Squares" since the removal had been so "vivement réclamée" by the population of the neighborhood.[44] The mères de famille in the area were more interested in the function and safety of their square than they were in an aesthetically pleasing promenade.

By March of the following year, news of the intended removal of the fountain had reached the business owners and residents with property adjacent to the square. Some on the rue Rochambeau and rue Lafayette submitted a petition to Adolphe Alphand. In it, they protested the destruction of the cascade and pond. Identifying themselves as "riverains," perhaps to underscore their proximity to the space, the petitioners claimed that the decision would "have no other result than the disadvantageous transformation of the square on the pretext that the pond exhales harmful miasmas that are detrimental to the health of area residents."[45] They concluded with the statement: "The undersigned certify that contrary to being an inconvenience, the pool and the cascade are in fact a source of purification and refreshment of the air for the *quartier*, and that the removal would be, in their eyes, an act of vandalism, no less contrary to the healthy aspirations of art than to public health."[46] Just over a week later a second petition, signed again primarily by riverains but on rue Baudin (now rue P. Semard) side of the square, arrived at the office of the prefect. "The undersigned," they wrote, "request the conservation of the square *in its present state*. We have no doubt that once the interests of those they represent have been clarified to the members of the Municipal Council, they shall be happy to agree to go back on a regrettable decision. The removal of the charming fountain that decorates the square, one of the most successful

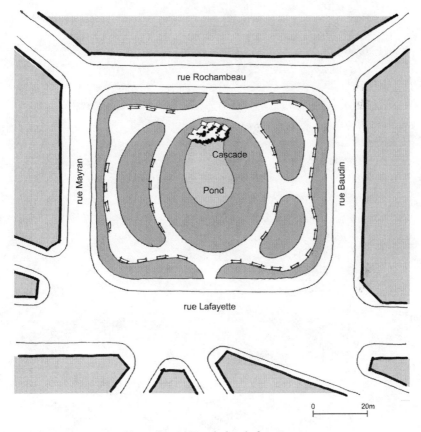

Map 5. Square Montholon, before 1874.

creations in all of modern Paris and admired by so many foreigners would not only be an offense to art but a grave loss to the neighboring homes."[47]

The riverains relied on the fact that their homes and businesses bordered the square, which they may have believed implied greater rights over the space. They invoked health, aesthetics, constituent rights, modernity, and national pride to combat the removal of the cascade and pond. Proximity, however, did not always determine the outcome in questions such as these when other public interests took precedence. At the Square Montholon, mothers constituted a significant percentage of park users, and their fears over the health and welfare of their children held more sway with the park service than the interests of the riverains. This primacy of children's interests in questions of public utility had ties to the treatises and publications of doctors and reformers in the 1850s and 1860s, which helped elevate children's health and welfare in the public eye, and was partly a result of growing concern in French society about depopulation. In many neighborhoods, the squares functioned as a kind of communal daycare center. This particular role of greenspace in the quartier translated into increased influence over the physical and social space by mothers for whom the square had become an extension of the domestic sphere. The park service removed the waterfall and pool; the rocky structure over which the water had cascaded was left in place. Engineers opened a large children's play area surrounded by benches in the center of the square.[48] (See map 6.) Thus, particular groups might link, either explicitly or implicitly, questions of

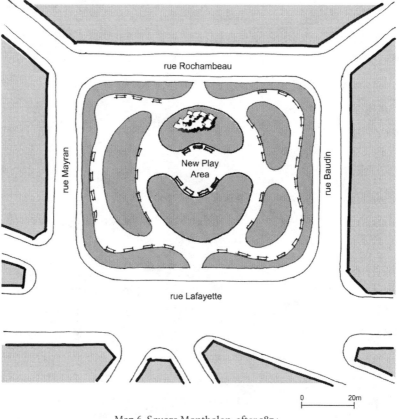

Map 6. Square Montholon, after 1874.

local usefulness to national and social concerns to gain greater control over the form and function of greenspace.

Indeed, park space was malleable, and many groups were not hesitant to request dramatic transformations. Residents of the Thirteenth and Fourteenth arrondissements signed a petition in 1895 calling for a redesign of the area around the music kiosk in the Parc Montsouris.[49] They wanted to make the space more conducive to their own rituals of communitas: the community concert. The group sent their petition to the office of the prefect in October of 1895. They did not receive a response quickly enough, so M. Tiger, who lived close to the park on rue Glacière and represented the petitioners, contacted the head of the park service directly in April 1896, no doubt to insure that the changes could be made in time for the upcoming summer concert season. The original petition was remarkably specific. The park users asked the service to eliminate two full lawns adjacent to the kiosk to make way for more seating close to the bandstand, and standing room behind. They also wanted two new trees planted in those open spaces to provide shade for concert goers. Finally, they wanted the park employees to "turn off" the waterfall during musical performances, ostensibly because it was too loud. This was a somewhat curious request since the cascade stood more than fifty meters from the kiosk, and the flow of water was steady but not strong. They included with their petition a landscape plan of the area on which they marked in watercolors the lawns to be removed and the placement of the new trees. In its report on the matter, the park service indicated that there appeared to be no reason why the petitioners could not be accommodated with the removal of the lawns. (Not surprising since maintaining grass in the squares had always presented expensive logistical problems.) The agency agreed to plant the two new trees. For these it suggested a chestnut and a catalpa, but indicated that it was already too late in the season to transplant them, so installation would have to wait until the fall. Finally, as for the waterfall, the park service issued orders to the cantonniers instructing them to shut the cascade down when necessary as requested.[50] The habitants of the Thirteenth and Fourteenth arrondissements had pressed for a redesign of the park space and a change in its operation to suit the needs of their community, and got everything they asked for.

This remarkable alteration to the layout and management of the park notwithstanding, actual design of the squares generally reflected the primacy of the interests of the children who played in those spaces. Alphand insisted that the squares should have "sufficient shade and free space for the play of children."[51] To achieve this end, in greenspaces such as the Square des Batignolles, Square Montrouge, Square Parmentier, and Square Delaborde, he widened footpaths at the corners, creating a kind of bulge or open space in the walk. Benches lined the exterior of these areas, and trees were planted liberally around them. Often, in squares such as those of Batignolles, Innocent, and Delaborde, for example, a single tree or a row of trees further delineated the play area from the rest of the path, at once creating shade and separating the children's games from the general circulation of park visitors.[52] Gravel paths suited popular children's games, such as the stick and hoop, jump rope, or bouncing a leather ball, much better than grass. Although grass became the preferred play space for children in the twentieth century, in nineteenth-century Paris, residents much preferred shaded gravel spaces surrounded by plantings. Indeed, Alphand's consideration of children's interest in the squares extended to the most minute detail of their play areas. In an 1869 report by the park service responding to a complaint about dust generated by paths that had not been wet down sufficiently during the heat of the summer at the Square Montholon, the service official suggested an alternative solution. He recommended using "river sand purged of the finest parts, that is to say *mignonette* [pea gravel]" to cover the paths, which would reduce the amount of watering necessary and keep down the dust during heat waves.[53] A note added at the end of the document by the official's superior indicated this was not a viable solution since a "memo from M. le Directeur [Alphand] had forbidden the use of *mignonette* because it can hurt children's feet."[54] Consideration of children's interest and use of the space sometimes helped determine details as seemingly insignificant as the nature of surface treatments.

Careful consideration of the neighborhood demographic, specifically the population of children in the surrounding quartier, weighed heavily in the balance when the park service made decisions concerning any proposed entertainments in the parks and squares. The service permitted concessions, if it felt they addressed the needs of the park goers and increased the utility of the park for the local population. In the predominantly working-class Belleville, the Parc des Buttes Chaumont drew large numbers of neighborhood children. One of the first authorizations of a new concession, beyond the original three restaurants, occurred in 1876 when the park service approved the operation of a small goat cart, which was pulled along a stationary track and could carry up to seven tiny passengers.⁵⁵ In 1889, the large numbers of mothers and children in the park led to the approval of the construction of a *Chalet Bébé,* a refreshment stand that would provide waffles, cakes, and sweet drinks at reasonable prices. In his report and recommendations on the question, Inspector Caillas of the park service noted that a great number of mères de famille frequented the park in the good weather, with children in tow. "The play, the movement, and the fresh air," wrote Caillas, "whet the children's appetite and the mothers are not able to procure anything for them in the streets of the neighborhood or within the park itself where there are three cafés that the parents cannot [afford to] take their children to." He recommended approval of the construction of the Chalet Bébé, calling it a "veritable service" to the public but stipulated that the price of all items sold at the concession should be limited to a maximum price of fifteen centimes per item. The park service quickly approved and established the chalet following Caillas's guidelines.⁵⁶ Subsequent installations with the neighborhood demographic in mind included a marionette theater in 1892 and swing sets and a merry-go-round in 1893.⁵⁷ When each of these concessions was granted, however, the park service attached similar requirements that demonstrated sensitivity to the economic realities of the lives of park users and the neighborhood. The proprietor of the marionette theater, for example, received his permit with the condition that he periodically put on shows without charge for students from local schools who would be brought there on outing days.⁵⁸ Concessionaires who agreed to these kinds of stipulations often lived in the neighborhood and were part of the community that pressed for parks and squares that were responsive to the needs of the quartier.

Given the numbers of children at play in the squares and parks, and considering the indulgence shown them, it is hardly surprising that there would be many complaints about misbehavior and damage. In 1859, a brigadier guard's daily report recorded that, in the Square de l'Archevêché, which was situated behind the Cathedral of Notre Dame, "between eight and nine o'clock, while the guard Langlois was making his rounds, ten or twelve children scaled the fence in order to throw rocks at the trees and obtain the chestnuts."⁵⁹ Many of the smaller squares did not have a permanently assigned guard. Rather, guards from squares nearby made rounds to the location to look things over. The Square de l'Archevêché was one such square, and its surveillance was the responsibility of the guard on duty at the Square Monge by the *École Polytechnique,* a good ten-minute walk away. On one of his tours of the square, the guard from Square Monge noted that the nightly raids of the children, who were either children of the streets or from the nearby neighborhoods, had significantly damaged the fountain in the Square de l'Archevêché. The fountain is in a very bad state, he wrote; "there are large rocks in the basin, and the sculptures are destroyed, the fins on the dolphin that spits water from its mouth are broken; if this continues just a bit more, the fountain will be totally destroyed."[60]

In the square of the Mairie of the Twelfth Arrondissment, which occupied the triangular space at the intersection of the avenue Daumesnil and rue Charenton, the damage caused by children resulted in a significant alteration of the interior space of the park and a rare case of reduction to the children's play area. The grass and flower beds in the center of the square had been removed at some point between its construction in 1878 and 1900 and replaced by a sandy play area similar to what had been effected in Square Montholon. In 1899, the city installed a bronze statue by the sculptor Ernest-Eugène Hiolle atop a large stone pedestal in the middle of that play area. Around its base was a low, looping iron fence. Since the square lacked any guard on duty, a certain M. Hiolle (perhaps a relative of the sculptor, who died in 1889) called the city's attention to the damage that had been done to the statue and to the low fence, which was bent and twisted from children balancing on it. He asked the park service to construct an area of grass a few meters across around the base of the statue and surround it with another, higher fence. This would, he believed, "prevent the *gamins* from completely deteriorating the work."[61] The park service agreed and completed the alteration. From grass to sand and back to grass—the park service could easily refashion the square since these inherently malleable greenspaces could be altered to fit current needs, whether they were those of children or those of the municipal administration. Although city policemen certainly tried to assist the park service by chasing off vandals or children damaging park property, this was never a truly viable solution, and a persistent lack of manpower often hindered the efforts of the Service des Promenades et Plantations to sufficiently protect public greenspace from damage, a fact that exasperated officials of the service.

Even so, the presence of guards in the squares was not a sure-fire defense against damage children might do, nor could it guarantee the children's protection from any number of hazards. Moral corruption was, for some of the Parisian mères de famille, a very real risk to their children, particularly when they visited the pleasure grounds of the *ancien régime,* which were now children's playgrounds. Statuary, common in the formal royal gardens of the Tuileries and Luxembourg Gardens, fit the *jardin français* style of those spaces, but was a constant source of consternation for many mothers. In 1861, after having been renovated and moved slightly to the side, the Medici Fountain in one corner of the Luxembourg garden was unveiled, and many mères de famille were shocked to see the figures designed by Auguste Ottin that now graced the front of the fountain. The arrangement told the story of Acis and Galatea. The lovers were portrayed completely nude save for a strategically positioned bit of fabric, and they lie recumbent in what appeared to be a rather amorous and suggestive embrace while the jealous cyclops, Polyphemus, hovered in all of his naked glory on the rock ledge above. The sculpture captured the moment just before Polyphemus hurled a boulder at the couple, killing Acis.[62] Henri Dabot, a young lawyer living in the area, delighted in the way in which the children "overwhelmed" their parents with uncomfortable questions. "The mothers," he wrote, "were not at all happy."[63] Many were already quite upset about the two nude "great devils of men in marble" which flanked a new gate and likewise drew the attention and questions of children. So shocking were the male statues, apparently, that the landscape crews in the garden, in deference to the mères de famille, refrained from trimming the shrubs around the bases so that the greenery might grow tall enough to screen the offensive bits.[64]

Beyond the potential for this kind of scandal in the redesigned royal parks (amusing to some, like Dabot) and despite the keen focus on children's interests in the design and management of the parks, children could often be in danger of much greater harm or injury in these spaces. They were at risk of incidents that ranged from the fairly banal dog-bite or broken bone to a frightening encounter with a child molester. On a spring day in 1899 in the Parc des Buttes Chaumont, for example, the park guard Dupeche witnessed forty-four-year-old Daniel Robert, who was seated on a bench, expose himself to a young girl nearby. Robert was arrested, charged with *Outrage public à la pudeur* (public indecency), convicted, and sentenced to four months in jail and a fine.[65] One month later, in the same area of the park, a twelve-year-old girl named Antoinette (Laboi?) encountered François Simard, age fifty-seven, just outside a park restroom by the café there. Simard took the young girl by the hand and attempted to force her to masturbate him. She escaped, and the man was apprehended. Like Robert, Simard was taken to the police station in the neighborhood and charged with public indecency.[66] The number of incidents of this kind in the Buttes Chaumont and elsewhere that are recorded in police registers suggests that the parks and squares with their large number of children at play, and sometimes unattended, attracted child predators. Guards in the greenspaces acted promptly if they witnessed or learned of an attack, and they did what they could to protect children from these and other kinds of dangers. They often took personal interest in seeing to it that a lost child was reunited with a parent, that bullying did not occur, or that an injured child received medical attention. Mothers often assisted the guards in helping to keep a close eye on their own children and those of others in the greenspace.[67] The squares in the quartiers thus served as nexuses of community linking neighbors even more closely to one another, guards to habitants, and the city administration to city residents.

There were other social groups that inhabited the parks and squares that were similarly less welcome in Parisian greenspaces than were the area residents, and mères de famille and their children. They were those who had slipped through cracks in society, or who chose to absent themselves from convention. They were youth gangs, vandals, the indigent and homeless, and prostitutes. There are few records that give voice to how these Parisians inhabited the space. The reports that do exist, however, shed light on the attitude of the municipal administration and habitants toward these "unwanted" park users. Throughout the last decades of the nineteenth century, much of the mischief that occurred in the parks and squares was the work of young neighborhood ruffians and their gangs who transformed the parks and squares into their own local hangouts, damaging property and harassing the guards and other park goers. The old guard at the Place des Vosges, M. Gras, had a particularly difficult time with the teenage boys who loitered about his square. In July 1900, Gras noted in his daily report that he had confronted one of the boys in the square during a summer concert. Just as the music began, M. Suard, who worked setting up the chairs for the performances, saw several young men climbing the posts of the bandstand and upsetting the people who were trying to hear the music. He scolded them to get down, and all but one did. Flustered with the young man's impudence, M. Suard called for the guard to take care of the situation. Gras ordered the boy to get down, but the young man refused, saying that it would take a "ministerial order" to make him leave his place. Articulating his own sense of proprietorship over the space, he told them that "neither the guard nor any employee could make him leave because he paid his taxes just like everyone else and he had the right to put himself wherever he pleased."⁶⁸ At this the guard used force to pull the young man down off the music kiosk. The boy began to shout at Gras and Suard, saying that his father was a journalist and could put whatever he wanted in his newspaper, and that they had no idea with whom they were dealing. He told them that "neither the police commissioners, nor the agents there, nor anyone else had any right to tell him anything."⁶⁹ The guard took the young troublemaker to the police, where he learned that the boy was Émile Brunneau. He worked for a paper company and lived on the rue Saint Maur. Gras left Brunneau in police custody and returned to the concert to continue his shift. Gaps in police records at the local level make discovering what happened to young Brunneau difficult.⁷⁰ What is clear is that these kinds of trips to the police station meant that even squares with an assigned guard might be without any guard presence for significant amounts of time throughout the day.

There were petty criminals and hoods in the parks as well, whose activities were of a more nefarious and opportunistic nature than the youth gangs who challenged the guards and disrupted the quiet of the neighborhoods. Thieves were particularly attracted to poorly lit neighborhood squares. They, too, were among the undesirable and unwanted park users. On 7 February 1888, M. Lhuiller and four other residents living near the Square Lamartine in the Sixteenth Arrondissement signed a letter to the prefect of the Seine complaining that there had been a recent burglary in one of the homes bordering the square, which they believed was facilitated by the nature of the landscaping in the park. Crews from the park service had planted a clump of bushes in the remains of an old artesian spring in the square. This now overgrown spot had become a hiding place, he claimed, for "prowlers," and he suggested that the shrubs be removed and replaced by a climbing vine, which would still mask the old spring.⁷¹ The lead gardener who responded to the complaint wrote that indeed "[t]he mass of green shrubs planted in December 1884 in the basin of the spring form today a compact mass of verdure in the middle of which criminals can lie in wait, but this is the same problem that exists in many places in Parisian parks." Rather than removing the shrubs, the gardener suggested placing a 1.5-meter latticework just behind the first row of bushes to prevent the criminals from secreting themselves in that place. Alphand approved the plan. The somewhat nonchalant statement about criminals hiding in squares and parks all over Paris suggests that this was an ongoing concern, and the solution adopted by the park service demonstrates that the agency never completely abandoned aesthetics in favor of a heavy-handed practical fix. Any design solutions to problems in these spaces had to blend engineering pragmatism and aesthetic sensibility.

Petty criminals not only preyed on the habitants, they also often targeted city property and the property of the concessionaires. Throughout the latter half of the nineteenth century, plant theft was a persistent problem for the park service. In 1862, the loss of plants from the Square des Innocents became the backdrop for an incident that demonstrated both the extent of the problem and the way in which the guards and city employees felt about their own proprietary sense of the space. In a letter to Alphand, one of the *Garde général des promenades intérieures* of the park service (a M. Bréton) explained that "[f]or some time now individuals enter the square of the Fountain of the Innocents nightly and steal plants. Since it was important to seize these delinquents and make an example, I instructed the guard Kohler to begin nightly surveillance."[72] Between midnight and one o'clock on the morning of May 4, Kohler noticed that a man had scaled the fence and was in the process of digging up the plants. Kohler gave chase. He eventually caught the man but only, according to Bréton, with the considerable assistance of a M. Vaquier, a gardener employed by the park service, and a M. Bacourt, a mason working the late shift for the sewer department. The thief's name was Tisset, and while the group waited for the municipal police to arrive, he reportedly offered to give his watch and chain to Bacourt in exchange for just letting him go free. Bacourt declined, and the police took Tisset into custody. Bréton asked that the three "brave" men be awarded a bonus of ten francs each for having successfully captured one of the thieves. Alphand agreed and added a fifteen-franc bonus for Bréton.[73] Why Tisset was stealing plants and their intended destination are unknown, but clearly the problem of plant theft had reached the point where extra measures and rewards were called for, as in this case. In fact, plant theft plagued the park service in all of the greenspaces in Paris into the twentieth century.[74]

Although plant theft was obviously unacceptable to the park service, it did allow the poor to gather resources like wood in the larger parks, with some limitations. In February of 1853, as work was beginning to renovate the Bois de Boulogne and in concert with that transition, the administration issued a directive that outlined new regulations beginning in March on the gathering of firewood by the indigent.[75] The order instructed the guards that gathering of wood would only be allowed on designated weekdays from 9 a.m. to 4 p.m.; anyone caught collecting wood at other times would be considered an offender. Using a tool to cut or break wood was forbidden. Collected wood could only be for family use and not for sale, and those who gathered wood larger than twenty centimeters in circumference would lose their privileges.[76] The city considered fallen branches and kindling a resource that it had the right to disperse as it saw fit, in this case as charity. In fact, on 22 October 1855, an *ordre du jour* from Conservator Pissot reiterated to the guards the legal reality that everything in the park was the property of the municipal government. In a near rant, Pissot admonished the guards for not doing their jobs properly. He told them that he had discovered that construction workers' wives and even the construction workers themselves passed daily out of the park gates taking with them large quantities of chestnuts and game birds that they had collected and poached: "That is what is going on and that is what you ignore. I tell you again, I am very unhappy about this! You are taking advantage of the fact that I am detained on other projects, but you might take care, the end of the year approaches and since self-esteem does not guide you, I will take you by your [self-]interest. You know well how to demand advancement, but you do not know how to earn it."[77] Pissot was responsible for all of the resources in the Bois de Boulogne and, for him, whether this situation represented carelessness or collusion was irrelevant (although Pissot seems to have believed there was a possibility of the latter). The administration took protecting parks from damage and against theft very seriously, as demonstrated in the prevalence of those kinds offenses listed in the *Registre des Procès verbaux des délits dans le Bois de Boulogne,* in which all kinds of park infractions from 1853 to 1880 were recorded.[78] Although some park resources were made available to the poor, workers, and the citizens of Paris, it was only at the behest of the administration and as controlled and managed by the park service.

Not only did the working or indigent poor use the parks' resources either illicitly or through the good auspices of the park service, vagabonds and criminals often inhabited the greenspaces. This was mostly true in the larger parks, where it was more likely that one might elude authorities. The guards in the Bois de Boulogne and the Bois de Vincennes during the Second Empire regularly ran patrols to round up those it identified as vagabonds and criminals who camped in the park at night and turned them over to police custody.[79] Smaller parks such as the Parc des Buttes Chaumont and Parc Montsouris did not have regular nightly surveillance patrols, but they likely had a nocturnal criminal element as well. Park service officials often expressed dismay at their inability to protect people from the potential for crime in these parks due to insufficient illumination and surveillance at night.[80] The squares had their share of those living on the margins of society as well, although the problem was so much less acute, and at times the park service chose to do nothing about it. The homeless in the area around Square de la Chapelle, located in the Eighteenth Arrondissement, lingered about particularly in winter since there was a *refuge de chauffoir* (warming shelter) on the boulevard nearby.[81] In the early spring of 1886, the problem for the park service was not that the homeless were present in the square, it was what they were doing while there and the damage the square suffered. In a report dated 31 March, Inspector Caillas of the service reported that each morning when the guard of the square reported to duty he found human excrement on the paths and the lawns. Caillas speculated that, since there were no toilets on the street next to the nearby shelter, homeless individuals were climbing the fence and relieving themselves in the park. He recommended that the city install a public lavatory on the boulevard to alleviate the problem.[82] In a direct response to Caillas's report, an engineer for the park service wrote that there already were facilities on the other side of the Place de la Chapelle near the bus stop, only ninety-seven meters from the shelter and open until midnight. He did not feel that even keeping that toilet open all night would solve the problem, let alone opening a new one, because the crowds that came to the shelter were simply too great. "The removal of the kind of heating apparatus in question," he wrote, "that took place on the first of April this year, will put an end to the issue of the gathering of homeless individuals at the spot, thus the inconveniences that have resulted concerning the cleanliness of the square ought to disappear."[83] The engineer concluded that no action was necessary at that point.

The presence of vagabonds in the squares could cause tension between local residents and the park service regarding what ought to be done. In 1900, M. Blondel, who lived at 3 rue Mayran next to the Square Montholon, wrote to complain about the "undesirable population" that "hog" the benches in the square, asking the park service to take some action to remedy the situation.[84] The conservator, under whose jurisdiction the square fell, responded to the complaint. "In this square," he wrote, "as in fact, in all the other parks, there are each morning a certain number of persons whose appearance leaves much to be desired, but who present nothing unseemly or any behavior that would motivate expulsion."[85] The guards normally roused "drunks and sleepers" if they found them among this group, he said, and they should be advised to redouble their efforts in that sense. However, the park service should not chastise the guards for the situation about which M. Blondel complained because it could incite them to "take clumsy measures toward persons who have committed no other fault than that of not being well-dressed."[86] Although M. Blondel wanted something done about those he considered vagabonds in the park, the park service official accepted the situation. His instructions to the guards distinguished between problematic behavior and mere presence in the square. This distinction reaffirmed that greenspace was *public* and by nature democratic, and only disruptive or criminal behavior, damage to property, or hindering another's use of the space in a significant way could deprive someone of the right to be there.

The poor and homeless were not the only inhabitants of the parks and squares whose presence caused conflict or tension; prostitutes (or those suspected of solicitation) often occupied the space as well and, depending on the kind of prostitution in which they engaged, they might not be welcome. Guards in the Bois de Vincennes and the Bois de Boulogne often apprehended women whom they identified as prostitutes. In August of 1854, M. Aubert, who was the guard of the Porte de la Muette at the Bois de Boulogne, recorded a *procès verbal* in his log concerning two young women in the park. The two were Jeanne Marie Chenu, age nineteen, and Louise Payes, age twenty-two. They had rented horses and intended to enjoy the warm weather by riding in the bois that day. Aubert took notice of them because they were riding "extravagantly," he wrote, "in such a manner that their dresses were pushed up to and *above* the knee"[87] They appeared to be rather drunk when they were stopped and escorted to the commissioner of police in the village of Neuilly. They lived, they said, in a *maison de tolérance* (licensed brothel) at 7 rue des Ciseaux, which confirmed for the guards that they were, indeed, prostitutes. From the beginning of the nineteenth century through to the first decade of the Third Republic, successive city administrations actively pursued a "regulationist" policy toward prostitution.[88] The logic of the approach was that, since prostitution was a kind of necessary evil, and because left unsupervised it presented significant moral and physical dangers to society, the city should regulate it closely.

The maison de tolérance was a crucial tool in that system. Brothels were licensed in only specific quartiers, and those houses were meant to be "enclosed worlds" wherein prostitution could be at once monitored by the state and rendered invisible to the 'respectable' public. Women who worked in the maison were required to obtain the permission of the brothel owner to leave the premises, and it was rare, indeed, that they received it.[89] Thus, on that summer day in the bois, young Jeanne Marie and Louise were unwelcome in the park, not because they were actively engaged in solicitation, but because they were outside the confines of the maison de tolérance and state supervision, circulating in public, far from the Sixth Arrondissement where the brothel was located, and engaging in behavior that made them obviously recognizable as prostitutes. Leaving the two with the police, who no doubt conveyed them back to the brothel, Aubert returned to the park. There he confronted the man who had rented the women the horses because, in addition to the concern caused by these young women appearing in public, it was strictly against park regulations to rent horses or carriages to persons who had been drinking. The vendor denied that he had committed any breach of the rules since he claimed he had not noticed that the women were inebriated, and he vehemently protested his innocence to protect his concession from revocation. Had the women been a different class of prostitute conducting themselves in a "decent" manner—a *cocotte,* or *grande horizontale*—they may not have gone unnoticed, but they would not have been ejected from the park.[90]

The presence of prostitutes, or solicitation, was not limited to the larger parks. In the squares of Paris, *filles de joie* enjoyed the greenspace in certain quartiers as much as anyone else, something that often rankled the neighborhood residents and particularly the mères de famille who also frequented the squares. The problem often was, however, sorting out whether the women were in fact prostitutes or just perceived as such, and if they were indeed engaging in solicitation. Mme. Amélie Larcher wrote to the Service des Promenades et Plantations in 1896 to complain about the lax way in which the Square de la Chapelle in the Eighteenth Arrondissement was policed. Parents no longer dare to send their children to the square because it is poorly kept, she claimed, and "at nightfall, *filles de joie* make a home there and give themselves over to their commerce assured of not being disturbed in their work.... Everyone is scandalized, but no one takes the trouble to complain."[91] The official who investigated the complaint found that "The complaints of Mme.

Larcher do not stand up." The square is not poorly managed, he asserted, and "the *filles de joie* of whom the complainant speaks, exist no where except in her own imagination."[92] In this instance, he recommended that nothing should or could be done, since Mme. Larcher imagined the problem—and did not include her address. The guards and the inspector having not witnessed any prostitution dismissed the complaint as being groundless.

However, the officials of the park service may have been quite incorrect in their estimation. Following the shift in attitudes toward regulated prostitution during the early years of the Third Republic, identified by historian Alain Corbin, public opinion turned against the idea of the maison de tolérance. Consequently, these establishments were less protected by the city and the police in the latter part of the century, and more subject to competition from various other forms of unregulated solicitation.[93] It appears that, in the quartier surrounding the Square de la Chapelle, this was particularly true. In a complaint to the city in 1893, an owner of a licensed brothel at 106 boulevard de la Chapelle complained to the city administration that his business was being threatened by "street women" soliciting in front of his establishment. He added that there were wine shops and hotels nearby at addresses 104, 114, and 116, in which prostitutes operated freely outside any regulation.[94] The brothel and the other businesses mentioned were all located but a short five-minute walk from the square, so it is quite plausible that Mme. Larcher's observations were not a product of her imagination.

Unwelcome, although sometimes tolerated, the homeless, prostitutes, petty thieves, ruffians, and youth gangs all exercised their right of access to, and use of, the spaces of the parks and squares in Paris, as best fit their needs and wants. In some respects, they took the park service at its word that it had created these greenspaces for the benefit of all Parisians. Hidden from view in other spaces of the city, they were visible at times within the spaces of the parks and squares. In many cases, and particularly toward the end of the century, the guards and the park service were powerless to eradicate them from the parks and squares or even control their use of the space. These groups were part of the social fabric of the city and of the neighborhoods, and they believed the squares were as much their space as that of anyone else. They too pressed their rights over the space. These spaces functioned as centers of communitas even for the proscribed of society. Public parks and squares, no matter how much local residents and authorities attempted to control them, were never sealed environments. Rather they reflected the diversity of the surrounding population, and they demonstrated the inherently popular character of Parisian public urban greenspaces.

Parisian citizens had always been the express beneficiaries of the public greenspace development program begun during the Second Empire and implemented by the Service des Promenades et Plantations. Leaders and administrators had repeated that claim throughout the nineteenth century. Although the parks and squares offered other payoffs, they had been, at their core, created for people, for Parisians. They were the target audience, the end users, and none more so than the mothers and the children who came to inhabit the spaces of the parks and squares. Neighborhood residents understood their place in the process of greenspace development, as evidenced in the sometimes bold demands and assumptions they made concerning the design, management, and use of the parks and squares. Indeed, even groups that were unwanted in the greenspaces—such as gangs, vandals, and prostitutes—shared similarly enjoyed rights to use. Each of these groups coalesced around collective interests in the space as either healthy places, ways to go to work, refuges, territories, or centers of community. In this way, Parisian greenspaces became catalysts for Turner's spontaneous communitas. At times, these groups clashed, but none wholly surrendered its own right to use the space.

The city parks and squares were public trust property managed by a city agency dominated by positivists and engineers who were interested in functionality and perceived success in sustainable solutions. This, coupled with the public's sense of proprietorship and willingness to press its needs, and the spatial practices of park goers (either overt or covert) created a dynamic and fluid generative process that increased over time and contributed to furthering the democratization of these urban spaces. This often dramatic effect over space by users was not possible in the same way in the commercial or residential spaces of the city where actual, legal ownership and self-interest inevitably trumped practices or any perceptions of proprietorship. Moreover, whether explicit or not, the existential, organized communitas that greenspace had engendered often couched its demands and approaches to the spaces in a profound sense of *rights* over them. This idea of citizens' rights, rooted in French national identity, had been a powerful rationale for the initiation of greenspace creation in the city, and now the thread had woven its way through to the park visitors themselves. Acting in certainty of those rights, even in the absence of any actual rights, habitants, mothers, children, and even those on the fringes of "proper" society inhabited the parks and squares. They joined with those whose livelihoods depended on them and those who had conceived of them to become instrumental in engineering a different understanding of this form of urban space, and indeed urban living.

CULTIVATING BROADER COMMUNITIES

•••• ••••

This [Luxembourg Garden] will always be the garden beloved by college students, working girls, old men, dreamers, and those who love flowers, trees, shadow and silence; silence that each Thursday middle school and elementary school children from the neighborhood, and every Sunday players of kickball and tennis, come to disturb.
—ALFRED DELVAU, *LES PLAISIRS DE PARIS*, 1867

•••• ••••

Just as greenspaces were emerging in every part of Paris as an integral part of urban life, the city was experiencing dramatic population growth. Industrialization, new kinds of jobs, and changes to the city environment made urban living increasingly attractive and generally less toxic than during the first half of the century. A period of massive urbanization expanded the capital's population in the latter half of the nineteenth century as never before and altered the spaces of the city. Such demographic changes also involved a reconfiguration of work and leisure time. In the earliest decades of the nineteenth century, Parisian pleasure gardens like Tivoli, Beaujon, Idalie, and la Grande-Chaumière had been the haunt of the comfortable and idle. However, toward the mid-century, alterations in work patterns and improved standards of living meant that Parisians from all levels of society increasingly found time to enjoy the pleasures of a visit to their own neighborhood squares or a park, often on a Sunday along with other city residents.

Georges Seurat's representation of an afternoon on the Île de la Grande Jatte stands as testament to the appeal of greenspaces in and around the metropolis, and the entertainments they offered. The island was located just outside city limits and the Parisian park system, yet it is emblematic of the relationship between these kinds of public spaces and city residents. The painting depicts a Sunday afternoon in a park on an island in the Seine. Critics at the time and art historians today have disagreed about the meaning of the painting and what Seurat was trying to convey, if anything, about his sociopolitical view of French society, or his interest in the science of color.[1] On a very superficial level at least, the painting places Parisians *within* greenspace. The figures may appear stiff or disengaged to some viewers, yet undoubtedly they populate, they inhabit, this park at a time in French history when greenspaces had been, and continued to be, engineered throughout the capital. Parks and squares, initially created for moments of repose, a leisurely stroll, or children's play became attractive to an increasingly varied set of park goers. As part of the growth of the urban population, new groups materialized in Paris who sought to engage in sporting and leisure activities in the parks and squares of the city. Those drawn to greenspace for the purpose of skating and fishing, as well as those who participated in organized sports such as racing, cricket, and cycling, all found the parks to be essential centers of their own brand of *communitas*, creating their own citywide communities rooted in the shared experience of an activity, expanding the social diversity within urban greenspaces, and increasing anxieties for some elites.

Just as neighborhood residents had pressed their interests over greenspaces in Paris, so too did these new, broader contingents come together to exercise their rights over the space. Thus, public urban greenspace in Paris had to accommodate a still greater multiplicity of uses to remain open to all and relevant to the lives of the majority of city dwellers. Inevitable conflicts arose and required negotiations between park users, the *Service des Promenades et Plantations* always functioning as chief arbiter with ultimate authority. Many Parisians, however, found common ground in advocating for alterations to the parks and squares that would suit their own particular leisure tastes. By the twentieth century, those groups of park goers, who had emerged around specific activities in particular parks, came to see their interest in the protection and preservation of all urban greenspace as something shared with other contingents. Thus, greenspace came to be a nexus for the cultivation of broader communities beyond the quartiers and despite persistent social divisions. The popularity of sporting events and mass entertainments in the latter half of the nineteenth century and the increased organization of those interested parties led to still greater public participation in the ongoing reconfiguration of the design and use of Parisian urban greenspace, and a reevaluation of the place of these spaces in the lives of ordinary citizens.

Although children were not a primary focus of the redesign of the Bois de Boulogne, as they were in the many smaller greenspaces throughout the city, and although the physical area of the park was much greater, the rich, propertied Parisians living in the arrondissements of the western *rive droite* perceived of the park as a kind of immense neighborhood square. Just as the squares in the quartiers functioned as centers of communitas, the Bois de Boulogne became an essential gathering place for Parisian high society. With the initial connivance of the government, the *haute bourgeoisie* laid claim to the large park and developed spaces of exclusivity and rituals of communitas that sought to reinforce that claim. Since the earliest days of the refurbishment of the bois in 1854, spaces such as those of the central lakes, the racetrack at Longchamps, and the *Cercle des patineurs* set the park on a trajectory to secure a position among the greenspaces in Paris as an enclave of exclusivity. That exclusivity was nowhere more evident than in the parade of wealthy Parisians in their carriages who circled the central lakes in the afternoon in order to see and be seen, a spectacle that came to be known as the *tour des lacs.* The physical center of the tour des lacs was comprised of the two large lakes near the eastern edge of the park. The reconstruction of the lakes in 1854 constituted a successful Parisian debut for a young engineer, Adolphe Alphand.

The lakes originated as a single body of water designed by gardener Louis Varé. Varé's original concept, however, was a dismal failure. The water that filled the single, large lake repeatedly leeched out through the sandy soil and left a swampy, stagnant mess in its wake. The grading was flawed, as well: too steep on one end, too low on the other, which added to the fiasco. The challenges of the task exceeded Varé's capabilities. Frustrated with the delays caused by the inability of the lake to retain water and disappointed in the gardener, who lacked engineering experience, the new prefect encouraged the emperor to replace Varé with Alphand. Alphand's solution was to design two lakes instead of one and employ terracing to account for the varied topography. He lined both of the lakes with a form of concrete to prevent seepage, and constructed concrete banks for permanent shoring.[2] He encircled the lakes with a carriage path in order to provide multiple observation points and sweeping vistas of the islands of this central water feature. The result was a new, modern (particularly in light of the extensive use of engineering) French interpretation of the *jardin anglais* style of landscape design.

On a visit to Paris in the 1860s, U.S. Senator John W. Forney noted that the lakes were a "lovely alternation of wood and water, promenade and drive."[3] Alphand had applied the science of engineering to the art of creating a tableau in landscape and had not sacrificed aesthetics. His circular paths echoed Gabriel Thouin's eighteenth-century designs that had facilitated the country gentleman's ability to tour his private estate, taking it in and taking ownership of it.[4] And yet these paths, and those throughout all the parks and squares in Paris, were for the public, for Parisians. City residents, each in his or her own way, would lay claim to the spaces as they inhabited them. The creation of the Grand and Petit lacs in the Bois de Boulogne, one of the earliest and most successful embellishments of the park, demonstrated the way in which spontaneous rituals of communitas could define a space and help develop a sense of proprietorship.

The significance of the lakes was, thus, not the ingenious design or the natural beauty which they recreated and celebrated. Rather, the area became significant through use; the tour des lacs became the most important function of this site. Each day at about four o'clock in the afternoon throngs of wealthy Parisians descended on the Bois de Boulogne in carriages and made several passes on the road around the lake before heading back to their homes in the city. Alfred Delvau described the tour in his 1867 Paris guidebook. "[G]oing to the Bois," he wrote, "is a tradition which one was careful never to miss . . . an excellent occasion to display one's horses or mistress, if one is a man, or to exhibit her *toilette* and critique that of others, if one is a woman."[5] The astonishing display of luxury particularly impressed American visitors to Paris during the Second Empire. John Forney sketched the scene for his American readers. "Of the number of vehicles present," he wrote, "I can give you no estimate, save that they seemed to be miles in extent, while on both sides of the carriage-way rode horsemen and horsewomen, attired as *only* French people can dress."[6] A young, recently widowed American southerner, Mary Bouligny, wrote of her impression of the tour des lacs during her visit in 1867. "The display of coroneted coaches," she marveled, "with coachmen and footmen in tight knee breeches, flesh-colored hose, shoe-buckles and gay cockades, is quite bewildering; and so much is there of the aristocratic swell and dignity, that a plain American who witnesses the scene for the first time, sinks back in the hired vehicle, feeling his utter insignificance."[7] Even the emperor and the empress participated in the tour des lacs. They appeared often "with a military escort," Bouligny recounted, "in a coach drawn by four or six horses, two of which are mounted by postilions.... The opportunity to see their majesties is always favorable, as the main avenue during the afternoon is too crowded for fast driving."[8]

Years after the creation of the lakes and the popularization of the tour, the city created another site within the park, a site that elites likewise transformed and altered through use. In 1856, the city granted a fifty-year concession to *Société d'encouragement pour le amélioration des races de chevaux en France*, known as the Jockey Club, and Alphand ordered the construction of the Hippodrome and the Tribunes of Longchamps.[9] The racetrack and the grandstands were set to become a space of tremendous exclusivity deep within this public city park. The grass track and wooden structures were little more than the physical features of a space defined as exclusive by those who used it. Further, the entire facility functioned to provide an outlet for patriotic sentiment, to uphold class distinctions and social structure, and to provide a distraction for the rich and powerful lest they become involved in Orléanist or Legitimist anti-Bonapartist activity. Established in 1831, the club epitomized bloodline, wealth, and propriety. Literary critic Frédéric Loliée recalled, "If one posed the question, 'Is he a Jockey?' it was the equivalent of asking: Is he well born, rich and distinguished?"[10] As well bred as their horses, and as well mannered, they were extremely suspicious of outsiders. Club members occupied the tribune immediately to the left of the central emperor's pavilion, and the one to the right accommodated princes of the imperial family, ministers of state, and leaders of the army.[11] The tribunes at Longchamps portrayed a vision of France dear to many elites, where the Jockey Club members were equal to the most powerful men in the government of France. Through their attitude, their behavior, and even their seating arrangement, the members of the société reinforced the exclusivity of their club and its prominence in society onto the site at Longchamps through a marriage of configuration and use. Rich and chauvinistic, the men of the Jockey Club were intensely patriotic, particularly when it came to any contest with Great Britain. Horseracing now took the place of pitched battle with their longtime rival from across the channel. In 1863, when an English horse took the prize at the first Grand Prix, the loss was tantamount to a national crisis.[12] The following year, when the Comte de Lagrange's horse, *Gladiateur,* regained the prize for France, the emperor dramatically broke with custom, descended to the field, and rewarded the Jockey Club member on the spot, naming him an officer in the prestigious Legion of Honor.[13] Practiced elitism on the part of the members of the Jockey Club at Longchamps, Alphand's early cooperation, and the emperor's patronage of their activities worked to uphold the exclusivity of the space.

On race days and Sundays, however, elite dominance of the space gave way to the whole of Paris. In the 1867 *Paris Guide,* Amédée Achard wrote that on race days, while the carriages of high society crowded the area of the racetrack, at the same time the "vulgar park visitors who come *en famille* to watch the spectacle of the Parisian races sit on the grassy slope above, which falls away to the cascade of the big lake. They are there by the thousands, with their wives and their children; artisans and pensioners, *petites bourgeoises* and *grisettes*—supposing there are still *grisettes*—everyone is there in a great jumble at once rural and democratic."[14] Édouard Goudon wrote in his popular study of life in the park titled *Le Bois de Boulogne: Histoire, types, moeurs* that on Sundays the crowds were "a motley crew, arriving from all points in Paris by foot or by milord; honest but little accustomed to good manners; good, but of mediocre dress; respectable, but pipe smokers; blessed by all the virtues of family but wearing ridiculous hats and dirty boots. They mingle with high society, but they don't join it."[15] These visitors were, in Goudon's mind, "the honest population of Sunday."[16] Longchamps, originally conceived as an enclave of the elite, could not entirely sustain its physical exclusivity in a public park. On Sundays and race days, the popular, public nature of Parisian greenspace forced elites to share their "square," like it or not, and some even took to avoiding the park altogether on those days.

For many of the haute bourgeoisie who perceived of the public space of the Bois de Boulogne as their own private space, this kind of mixing with "lessers" was troublesome, and they sought to create other places within the park where they might escape such proximity to the popular classes. The Cercle des patineurs was thus established in a secluded portion of the bois called *"la pelouse de Madrid,"* north of the tip of the *lac inférieur.* The skating club was not open to the public; rather, sole ownership and use of the area belonged to a small, private skating club of the same name. The emperor's favor assisted in the award of the concession to the group. Adolphe Alphand designed the rink and landscape for the members, and once again called on Parisian architect Gabriel Davioud to build the primary structure on the site, the Grand Chalet.[17] Work on the complex began in February of 1865, and quite remarkably, the Cercle des patineurs was available for use late that same year. The site boasted a shallow cement basin 250 meters long and 50 meters wide; a tall, wrought iron fence set back fifteen meters from the border of the basin; and a *"grand chalet en bois"* measuring some 32 meters long with a deck 9 meters deep.[18] It was the most modern and safest way to skate in the park—in winter when it was cold enough for the water to freeze, the shallowness of the basin made the water in it freeze faster and thicker than on a deeper lake.

Architect Gabriel Davioud patterned the Grand Chalet after a Swiss chalet and outfitted the interior to accommodate the most privileged of guests. The Grand Chalet consisted of a large central pavilion flanked by two smaller satellite pavilions and connected by galleries. Although built of common pine, the chalet's decoration was elaborate, with "lambrequins, finials and balustrades of denticulated wood."[19] The central pavilion, with its large cast iron stove in the middle of the floor, provided guests with amenities such as a cloakroom, a buffet, a lantern room, and toilets.[20] The pavilion to the left housed the Imperial Salon, and the one to the right, with its small, adjoining office, was reserved for the leaders of the club. Two galleries, each lined with benches surrounding a smaller, central cast iron stove, linked the three pavilions. Here the domestics and coachmen of the skaters warmed themselves during the winter parties.[21] The simplicity and unpretentiousness of the layout of the Cercle des patineurs belied its function during the reign of Napoleon III.

The emperor and empress as well as the members of the club hosted extravagant skating parties. Indeed, the Cercle des patineurs became an extension of the *Palais des Tuileries* in the Bois de Boulogne. Author Arsène Houssaye attended many of these occasions. Writing in 1885, he reminisced about those sparkling evenings and provided an insight into just how extraordinary the scene was. "One could imagine," he recalled, "that the Marshals' Salon on the day of a ball had been magically transported to the Bois de Boulogne. The Emperor and the Empress, all of their esteemed court; the ministers, the ambassadors, their coterie from the Tuileries and Compiègne would all relocate themselves there, at times beneath the fog of three o'clock, at times under the evening stars accompanied by paper lanterns."[22] The Cercle des patineurs, surrounded as it was by a fence and far from the larger paths in the bois, was one of the few completely private spaces in the park. *Le Monde illustré* openly criticized this exclusive enclave in 1865, writing, "in our opinion, the measure is perhaps too radical, and this exclusion *en bloc* of all those who are not members of the *Club* does not seem to us to be in accord with our egalitarian customs. We have heard murmurs on the edges of this privileged ice; and many wonder by what right the enjoyment of common property is the possession of few certain someones."[23] Such private spaces became increasingly rare as Alphand's influence grew and his own vision of greenspace as fully public manifested itself in the many decisions of the park service. Thus, in spite of these early attempts by elites fully to dominate the park space, the vast majority of the bois was and would remain a public park, open to all.

Greenspaces as popular spaces of leisure for more and more Parisians required ever-increasing management by the park service. Skating was an extremely popular activity for members of all classes in the Bois de Boulogne and the Bois de Vincennes. It captured the public's fancy, partly because the imperial family made the sport fashionable and were quite good at it, and partly because it required little or no special equipment, making it an activity that Parisians at most income levels could enjoy. Just one year after Alphand had fixed the problems plaguing the lakes in the Bois de Boulogne, Auguste Pissot, the conservator of the park, tried to develop some guidelines for skating there. Pissot called for limiting the activity to certain bodies of water in the park and only certain areas of the lakes since there were sections that never froze completely. He sought to ban sleds on the ice because their runners cut deeply and marred the surface. Since beverages were so expensive in the establishments around the lake, Pissot suggested granting four temporary concessions to poor women, permitting them to sell drinks to skaters on the ice of the lake. He believed the rental of skates should be limited to only a few licensed and regulated lessors to avoid "illicit commerce" and to control prices. Young people will wish to learn to skate, he noted, so they would do well to have a few qualified teachers approved by the park service. The service would be responsible for sweeping the snow off the ice and indicating when it was safe to skate. Finally, canes or other such sticks would be expressly forbidden on the ice because "one might become excited and start gesticulating wildly, perhaps striking a fellow skater, which could deliver serious wounds to the face."[24] Even while permitting the activity, the park service maintained tight control over it and the space.

In spite of Pissot's well-intentioned measures, some skaters, as expected, did indeed became overly enthusiastic and careless on the ice. This led to a terrible skating accident, which took place on a winter afternoon in 1862 and shook all of Paris. After eight days of freezing weather, the ice appeared to be quite thick on Sunday, 19 January, the day of the tragedy. *Le Monde illustré* reported that there were so many skaters on the lake that day that one could barely see the ice.[25] On the far side of the lake, the park service had cordoned off an area of particularly dangerous ice where the city had recently cut blocks destined for the newly constructed ice house in the bois. The new ice was not as thick or as safe as was the ice elsewhere on the lake. At about 3:15 p.m., the paper reported, when the crowd was at its largest and its enthusiasm most lively, a few skaters had inadvertently knocked down the pickets that held the cord delineating the thin ice. Soon, unsuspecting skaters moved on to the dangerous section, completely unaware of their precarious situation. One young man began to feel the ice giving way beneath his feet and tried to crawl on hands and knees to get back to the thicker section, but the ice collapsed under his weight and he plunged into the freezing water. A friend nearby tried to help the man, and he too fell through the ice into the frigid lake. Within moments, twelve people had fallen into the expanding hole in the ice. Bystanders tried every means possible to save the drowning skaters but were able to save only eight, who had survived by clinging to chunks of ice. There was no lifesaving equipment, and the boats by the side of the lake made for summertime rowing were useless in the rescue attempts. Onlookers eventually pulled three more skaters from the lake, but those victims were unconscious and died before reaching medical attention. In a gruesome twist, one body slipped under the ice and floated toward the edge of the lake. It remained there for another week before it could be removed.[26]

There were many reports of heroes that afternoon. The future father-in-law of young lawyer Henri Dabot was at the lake that day. The gentleman reportedly retrieved his "prohibited" cane from the lakeside and used it as a hook to save some of the people. Others in attendance described seeing Dabot's colleague M. Mulat leap into the frigid water and bravely save several people.[27] Details of the incident and engravings of the tragedy filled the pages of the popular press; Parisians all over the capital learned of the horrors of that afternoon. Indeed, the accident became such a popular measure of a man's bravery that there were more tales of heroes saving people than of people saved.[28]

In the wake of the tragedy, the park service began a vigorous investigation of the incident and explored means of preventing a reoccurrence. Within one month of the accident, Pissot produced a lengthy report in which he evaluated several potential lifesaving techniques, some of which were in use at the time in Great Britain. The systems included: ladders stationed around the lake, which when laid flat on the ice permitted the rescuers to advance on it without breaking it; small lightweight boats with blades mounted on the hulls that would allow them to be pushed easily across the ice to the location of an accident; lakeside stations with ropes and anchored hitch stakes; and cement pilings set at intervals in the water so that drowning persons could grab on and pull themselves up. Perhaps the strangest idea was one designed specifically for the Mare St. James, which was a public skating spot favored by the emperor. Pissot suggested suspending a massive mesh net beneath the ice that would catch skaters (including the emperor) should the ice break.[29] This particular installation was never put into place, but the ladders, ropes, and specially designed boats came into regular use in the parks.

Efforts to secure public safety went still further. In mid-February 1862, a directive signed by both Alphand and Pissot outlining the establishment of an ice-rescue team circulated among the guards and other employees of the park service. The order indicated that henceforth three small boats would be placed near the spots on the lake that were prone to a skating accident. The park service assigned four sailors to this special duty, and the boats were fitted with long rescue hooks and ropes. A fourth, smaller boat was placed near the Mare St. James with one sailor assigned to it. When the lakes were frozen, two regular guards were stationed around the Mare St. James and six guards around the larger central lakes. Alphand indicated that these special guards would be selected from among the most intelligent and energetic men in the park service corps. Pissot was in charge of executing the order and the selection.[30] Later that same year, Pissot submitted a memo to Alphand in which he reported that, given the increase in the size of the crowds skating in the bois, an additional twelve guards should be stationed around the lake in the coming season. The park service outfitted these guards with skates so that they could move quickly across the ice, both to assist in rescues and to prevent accidents by keeping skaters away from dangerous areas. Park laborers, employed to assist the guards, would augment their number on Sundays when the crowds on the lakes were the largest.[31] The park service was committed to keeping the public safe as Parisians engaged in the tremendously popular winter sport of skating on the lakes.

In November of 1862, just as the skating season was beginning again after the previous winter's tragedy, officials posted a public notice, signed by Alphand, at various points around the border of the lakeside. In it, he described the new precautions the park service had taken to prevent or deal with an accident like the one that had occurred in January. Alphand enumerated the administration's efforts to keep the public safe, and explained how lifesaving stations had been set up and supplied (with the ladders, boats, and ropes described in Pissot's February report). Park personnel would man these stations, and Alphand requested that the public kindly stay clear of any accidents and allow the park service employees to do their work. The disaster of January 1862 had shown administrators that failure to manage such activities carefully could have grave consequences. Throughout the summer of that year, park service officials worked tirelessly to set up new lifesaving systems, reassign personnel, and issue new regulations, with the goal of allowing the sport to continue while ensuring public safety.[32] Rather than banning the sport entirely, which they might well have done, given the shock of the accident, they chose instead to adjust their operations to allow for its safe execution. The following January of 1863, a plan was drawn up to make delineations on the lake which would further protect skaters. A watercolor of the plan shows that the park service designated the entire northern end of the lake as "*Dangereux*." The southern end was set aside for the sleds used to push women and older park visitors who preferred sitting to skating, and the central area designated for skaters. Just as the park service had delineated the carriage, bridal, and pedestrian paths in the park to keep circulation separated, it also divided traffic on the lake to prevent collisions and mishaps. The emperor approved the plan.[33] Thus, the public's desire to utilize the park in a particular way and concerns over safety altered the management of that space and increased the city's control over it.

Pissot's July 1862 report went beyond safety and included suggestions about other aspects of skating in the bois. The sport in the park had become so popular so quickly that the park service had not had time to work out all of the necessary alterations to the space to accommodate the public, leaving some questions open for consideration. In his report, Pissot recommended that benches should be set up around the lakeside for skaters to sit and put on their skates, and that the park service ought to establish approximately twenty new concessions to be granted to young men who would help lace skates for a small fee. The last paragraph of Pissot's long report hints at yet another attempt by some elites to shape the interior of the park to suit the needs of their own particular communitas. Pissot wrote that "many skaters, belonging the elevated classes of society, find it inappropriate for them to be mixed in with everyone else. . . . [I]n consequence, many have asked if it would be possible to reserve a section for their small number which would be always well managed. They have offered to pay a small fee [to the city]."[34] Pissot indicated that he saw no problem with such an arrangement. "This would bother no one, and give satisfaction to a good number."[35] In fact, he suggested that the park service could deposit the money in the mutual aid society for park workers, and in this way city employees would benefit as well. Although not a formal petition, this was nevertheless a clear effort to restyle the space of the lake to fit a particular need for exclusivity. Alphand had been head of the service for nearly a decade in 1863, and that which he might have assented to early in his career, he now rejected. He crossed out the last paragraph of the report, which concerned reserving a section on the lake for elites, and made a note concerning his deletion. Unfortunately, the note is illegible.

Nevertheless, it is plausible that there was some discussion about the matter between Pissot and Alphand in which Alphand expressed his displeasure at the idea because, in only a few months, Pissot's opinion on spatial social divisions appeared to have completely reversed. His response to a request in November of that same year seems oddly not his own in light of the earlier record. One entrepreneur, M. Borel, petitioned the park service for a monopoly on skate rentals at the lake, and he asked for permission to set up a tent for this purpose, which would include a section for society women to have their skates laced for them. According to Borel, this would fill a lack in the accommodations for skaters. "To begin with," responded Pissot, "I don't think the public has complained. There are plenty of skate rental locations where they can also receive assistance lacing their skates." Furthermore, Pissot continued, "[a]n ostentatious tent for one class of park users would have the appearance of establishing a distinction and would disgruntle others."[36] A few months after having advocated cordoning off a section of ice for those of the "elevated classes of society," Pissot here rejected Borel's request, arguing against a spatial distinction based on class. The reversal was complete and likely originated with his immediate superior, Alphand, since equal access to all citizens, and distribution of greenspace throughout the city, were increasingly recurrent themes in his directives and in his published works.[37] Elites might have had their tribunes at Longchamps, and later carved out the Cercle de patineurs, but the park lakes, as far as Alphand was concerned, would remain public space as egalitarian and as free of distinctions as possible.

By all accounts, the crowd of skaters in the Bois de Boulogne was, indeed, a diverse group. In 1862, *Le Monde illustré* recommended champagne to its readers as the perfect drink to accompany a skating party in the park, and suggested a *"costume obligé"* patterned after skaters on the plain of Madrid, where the Cercle des patineurs was located. Yet at the same time, the journal noted of the large numbers of working-class children who enjoyed the space as well: "they are children in *sabots* [wooden shoes], who cannot always buy skates yet who are neither less skilled nor do they enjoy the ice any less than any others."[38] Commenting again on the diversity of those engaged in the sport in 1865, the paper wrote, "It is not only the upper classes who give themselves over to skating; one recalls the ardor with which the shady world and even the bourgeois of Paris were given to this pleasure last year."[39] The lakes and ponds of the Bois de Boulogne, according to *Le Monde illustré*, "were neutral spots where all social positions rubbed elbows and nothing was more picturesque than this mixture and blurring of positions and races."[40]

The lakes proved to be truly public and communal greenspace when they were whitened by winter. The young American Mary Bouligny recalled that, following a cold snap in 1867, "the large lake at the Bois presented a grotesque sight, with representatives of every class engaged in the *glissant* motion. How animated was the scene, as we witnessed it from a little knoll overlooking the locked-up stream that shone like a mirror, and how excitedly the crowd skimmed along, marking their love of the novel amusement! Some one [sic] said within my hearing, that this was Paris on a Toodles—drunk; and so it seemed, as round and round they went,—some on their knees, and others lying at length, wholly unable to keep their feet."[41]

In fact, skating became so popular that, by the 1870s, the city's management of it necessarily expanded further still. In 1871, the prefect of the Seine, Léon Say, released a new set of statutes governing the practice of skating in the bois. In addition to skate vendors and renters, the city now granted concessions to skating instructors and lace-boys, as Pissot had recommended. All concessionaires had to carry color-coded identification badges so that the skaters could recognize them as legitimate.[42] Since lace-boys not only laced skates (they and the skate renters sometimes kept an eye on personal items such as shoes or canes), the badges could help prevent petty crime and theft by letting people know whom the city had vetted. Records of these identification cards and approved applications from 1871 to 1875 show that, in every case but one, the concessionaire resided in one of the villages surrounding the park such as Boulogne, Neuilly, Suresnes, and Saint Cloud, not in the affluent western arrondissements of Paris that also abutted the park.[43] The park thus served, to a certain extent, as a source of temporary or supplemental employment to residents of those particular nearby communes. There is no evidence that women ever obtained any of the concessions related to skating save for the few granted to beverage porters. Through the sport of skating, the Bois de Boulogne emerged as a communal space in which Parisians from all walks of life regularly engaged in an almost heady kind of existential communitas. In this way, a large greenspace shared much in common with the smaller squares when it came to functioning as a public space, not just for the residents of the immediate quartiers, which it certainly did, but also for the larger population of the city.

When the ice thawed, park waters naturally became enticing to anglers, but they remained frustrated by the city's ban on fishing in the parks. In mid-nineteenth-century Paris, the municipal government considered the resources of the parks to be city property, and harvesting them (poaching, fishing, and gathering without permission) or damaging them was a crime.[44] In the recommended guidelines for the management of the Bois de Boulogne, which Alphand submitted to Haussmann in 1856, fishing appeared in Article 8 along with trampling the lawn, hacking off the tops of trees, peeling bark, cutting down trees or shrubs, pulling up flowers, throwing stones or other objects in the streams—all activities that were "in a word damaging in any fashion that which exists [in the park]."[45] In 1857, Haussmann signed an additional ban on fishing along the banks of the Seine behind the newly constructed tribunes at Longchamps. The first article of the statute read: "As of 1 May 1857 it is forbidden to place along the length of the Seine bordering the plain of Longchamps keepnets, nets, and other fishing devices."[46] The statute was limited to fishing from shore at Longchamps and not the entire length of the park that bordered the Seine because the concern was the damage fishermen and their equipment might do to the newly installed landscape behind the racetrack, and not fishing in the Seine, which generally posed little threat to the park. However, Pissot suggested extending the ban on fishing to cover the full length of the park that bordered the Seine, again citing the potential for damage to the grass and plantings beside a new road that ran along the river in that section. Fishing remained illegal in the lakes and streams. Yet prohibitions such as these did little to properly manager the living resources of the parks and, throughout the 1860s, the park service assumed even greater wildlife management responsibilities.

By 1872, the park service decided to authorize fishing in the Bois de Boulogne and the Bois de Vincennes in a limited way, confining the practice to the smaller ponds in the Bois de Boulogne and three specific locations in the Bois de Vincennes.[47] Four years later, in 1876, Pierre Charles Dubois, living at 31 rue Doudeauville, in the working-class Nineteenth Arrondissement, led a group of men from the area surrounding the Parc des Buttes Chaumont in petitioning the city council to allow fishing in the lake of their local park as well. In the report, which resulted in approval of the measure, the park service official indicated that the presence and vigilance of the fishermen would assist the park guards in preventing damage to the plant and animal life within the park done by little urchins (*gamins*).[48] There was also some consideration of how much the area was still in the process of recovering from the upheaval of the siege and the Commune, something that park service officials were sensitive to and expressed in earlier renegotiations of contracts with concessionaires in that park. To grant fishing permissions would, in a small way, help local residents, who were still struggling in the wake of the conflict, to supplement family economies. It would also mirror existing regulations in the Bois de Boulogne and Bois de Vincennes, without setting a precedent. Records of the licenses granted to fishermen in the Parc des Buttes Chaumont from 1876 to 1892 show that the city granted as few as five and as many as sixty licenses each year, an average of thirty-four licenses per year.[49] Although the number of licenses varied, the fee remained constant at ten francs, approximately one and a half times the average daily wage for a male laborer. Detailed licensing records reveal a great deal about the fishermen Dubois had organized.

The group included men who lived near the park and renewed their licenses yearly, as was the case with M. Gilles of Boulevard de la Villette; and those who changed their residences but remained part of the group, as was true with M. Zubler, who maintained his license even as he moved about the vicinity of the park several times over the years from rue Doudeauville to avenue Laumière to rue Meynadier. Some of the fishermen were neighbors, residents of the same buildings, or even family members. The Lagolte brothers, Charles and Joseph of 94 rue d'Allemagne fished the lake as did MM Vallette (28 rue des Envierges) and Comte (92 rue d'Allemagne) and their sons. Although the majority of the men were from the surrounding area, concessionaires also joined the group of licensees, as was the case with M. Bouquet, who operated the restaurant at the Pavillon Puebla within the park, and whose business the city acknowledged had suffered greatly during the Commune. These men were so confident of their right to fish the lake that, not only did they secure the city's approval, in 1898 they petitioned the park service again to restock the lake with fish for their purposes.[50] The fishermen constituted a kind of informal fraternity of local residents brought together by a shared interest in fishing and a concern for ongoing use of the park space. The public park then functioned as a nexus where community and private family ties intersected, the park and neighborhood melded together, and public and private concerns coalesced.

Although the park service allowed licensed fisherman to operate in the Bois de Boulogne, Bois de Vincennes, and Parc des Buttes Chaumont, it continued to resist granting permission to fish in the Parc Montsouris, preferring to maintain that park as a hatchery to help maintain the fish populations in the other parks. In 1892, M. Kahn, living at 118 rue Monge, petitioned the park service for the right to fish in the lake in the Parc Montsouris because it was the lake nearest to his home.[51] The official who responded to the petition pointed out that the park service had never granted permissions to fish at this particular lake because the walkways around it were so close to the water's edge that casting might easily injure children and passersby. Moreover, the park service now used the lake as a fish hatchery to restock the lake at the Parc des Buttes Chaumont and the other parks. Finally, he noted that M. Kahn's address on rue Monge was no farther from the Seine than from the park and he could fish there, so denying the permit would not create an undue hardship for him by depriving him of access to a fishing spot.[52] Proximity in this case was a contributing factor with the primary reasoning being the park's configuration and its place in the larger system of parks. Two years earlier, the park service had likewise declined a request by M. Travet, who lived on the edge of the park. M. Travet wrote, saying that he recently noticed that there were a number of red mullet fish (a good food fish) in the lake, and that he would like to buy a half-dozen that he would fish out himself. The park service official who reported on the request indicated that it could not be approved because of the layout of the lake (the inconvenience fishing would cause to children and other visitors) but much more importantly because M. Travet had asked to *buy* the fish and selling directly from the park to a buyer would have set an undesirable precedent.[53] Harvesting by the city, charity, or granting permission to access park resources was an acceptable means of distributing those goods, but direct commerce was unacceptable.

When it came to fishing in the Montsouris lake, collective petitioning did not help. Anglers who organized themselves and petitioned as a group in 1897 to fish in the Parc Montsouris met with no greater success at obtaining the right than did M. Kahn and M. Travet. The park service rejected the petition, again citing the children and other park users, and the role that Montsouris played as a fish hatchery.[54] Fishing licenses could never be granted at Montsouris, in the view of the park service, because in addition to its function in the neighborhood, this park had a role in the sustainability of fish populations in the larger network of greenspace in Paris, which took precedence. This was not a question of the right of access, but rather a question of the parks as a coherent system.

Skating and fishing were leisure activities for which residents sought use of the park space; cricket, on the other hand, was an actual organized team sport that required the open greenspace that, in Paris, only a park could provide. It was a game popular among elites and foreigners; thus aficionados not surprisingly sought to obtain permission to create a permanent cricket field in the Bois de Boulogne located near many of the foreign embassies. By the 1860s, impromptu games hosted by cricket clubs were cropping up in the more out-of-the-way sections of the bois. In 1863, one of those clubs, the Paris Cricket-Club, wished to secure a specific location and provide some refreshment to players and spectators. The club's spokesman, Thomas Sumpter, a resident of the Hotel d'Albion, wrote to Baron Haussmann requesting permission to offer refreshments to club members who routinely held matches in the park. He also wanted to formalize the request so that his club would not have to obtain the same permission each week.[55] Haussmann acceded to Sumpter's request and allowed the club to play on the Plain of Madrid, near the Cercle des patineurs, with the understanding that these were to be brief morning games while the park was relatively empty.

A few months later, the park conservator, Pissot, submitted a report to the park service indicating that problems were beginning to emerge as a result of this arrangement.⁵⁶ He reported that, although the prefect was under the impression that the games were brief, in fact they often lasted well into the afternoon. Moreover, on Sundays, the club held its weekly play-off matches. On this day, the crowds were quite large and the refreshments considerable. The organizers set up a huge table, which Pissot indicated was one of "thirty to forty tablecloths," and the few refreshments were, in actuality, a "veritable feast."⁵⁷ Caterers who delivered the food and drink often drove their wagons across the lawns, creating ruts and other considerable damage; so much so that, on one occasion, an irate Pissot had to intervene personally to halt the damage. Worse yet, according to the permission Haussmann had granted the club, the refreshment table was supposed to be hidden as much as possible from view. But Pissot reported that this massive feast was not at all hidden amongst the trees; rather, it was in the open and easily visible from nearby carriage paths, making the Plain of Madrid appear, to Pissot, like one of the unseemly (in his view) *"guinguettes de barrière."*⁵⁸ "I understand that these gentlemen need their refreshments," wrote Pissot, "but then they might easily set up a tent in a nearby clump of trees, large enough to hold all of their necessities, which they could enjoy in the interior, so that one could not see what one does now—jugs placed on sawhorses and dirty tables like those in a wine shop. As for food after the match, they are close to the Restaurant de Madrid where they can find anything they desire."⁵⁹ Pissot was concerned again with damage to the park, just as he was in the case of the fishing along the Seine, but also with the experience of other park visitors who desired a picturesque carriage ride. There is no indication what action the park service ultimately took in this regard. However, subsequent similar requests by cricket clubs make no mention of an existing Paris Cricket-Club installation to plea their case, and it is rather unlikely that a tent or similar edifice was approved since the park service had so soundly rejected the idea of one on the lake in the winter of 1862. It is quite plausible that Haussmann's open approval came to be seen as a mistake. Pissot certainly thought it was, since it lacked the kinds of stipulations that were customarily included in approvals and concessions granted by the park service.

In 1878, a certain M. Kempfen pressed to establish a permanent cricket field in the Bois de Boulogne, although his affiliation is unclear. Alphand responded to the request in a memo to his chief engineer for that section of the city. In the memo, he indicated "there are good reasons why we should not cut off any part of the park currently available to all *promeneurs* and why we should not sacrifice any one lawn."[60] On the other hand, he felt that "an absolute prohibition of the game in the Bois de Boulogne might be seen by players of the sport as a vexing measure."[61] There was, however, according to Alphand, a middle ground. He determined to permit cricket teams to play on designated fields in the park only, areas that were conducive to the sport, yet would not be damaged by the players or the spectators. The park service rotated which spaces would be available, and the teams had to inquire of one of the guards in advance to know where games were authorized that day.[62] The order was passed on to Pissot, who was instructed to execute it in such a way as to allow the "greatest liberty to amateurs," while at the same time "disallowing any gathering or activity that would adversely affect the upkeep of the grounds."[63] Pissot would choose the location of the gaming fields using information on the condition of the grass from the men charged with maintaining the lawns, and the installations would be limited. Alphand's decision backed away from Haussmann's earlier unrestricted approval given to the Paris Cricket-Club. And yet, with this compromise, Alphand balanced access and use of the space with maintenance of it, considering all interests and privileging none.

Cycling was another sporting activity that became increasingly popular in the second half of the nineteenth century. Many Parisians wanted to use the spaces of the parks for their rides and races. Unlike skating, where the main concern was the safety of participants, or fishing, which was primarily a question of the management of municipal resources, or cricket, where balancing damage to the park and equitable enjoyment of the space were at issue, the complication for cycling was without question the disruption and danger it posed to other park visitors. In 1868, M. LaBrousse, a retired naval lieutenant, petitioned to hold yacht and bicycle races in the Parc des Buttes Chaumont. The park service rejected it, calling it "absolutely inadmissible."[64] LaBrousse had indicated that, during the races, spectators could be expected to line the lake and the paths of the park. According to the park service, this would prevent those who were visiting the park but not interested in the races from enjoying the space, and there was also the potential for damage. Worse still, the event would necessitate a fee to enter since the race course ran throughout the interior of the park. The Parc des Buttes Chaumont would, in effect, cease to be completely public. The park service official wrote, "It is clear that we can not abandon, even momentarily, the totality of public promenade to one particular enterprise."[65] He suggested that an event such as the accompanying proposed regatta would be more appropriate on a much larger park lake such as the one in the Bois de Vincennes, for example, where space might be set aside, and that the petitioner might coordinate something with the *Société des Régates*, a group that held races on a corner of the lac de Charenton without incident. As for the bicycle race, "because of the serious dangers it presents, it must be forbidden in an absolute way in any promenade where we must have in view, above all, the safety of the public."[66] Indeed, the city formally banned bicycles in all of the parks except the Bois de Boulogne and the Bois de Vincennes in 1871. Cyclists in those two parks were, nevertheless, restricted to carriage paths, required to have a bell on their bikes and a visible placard bearing their name and address, and were prohibited from cycling on horse-racing days due to the large crowds in attendance.[67]

As cycling grew in popularity, however, the city responded by relaxing its restrictions somewhat. A prefectorial statute in 1888 amended the law to allow cyclists to enter the Parc des Buttes Chaumont. Yet they, too, were to abide by the same regulations that existed in the larger parks.[68] Problems arose almost immediately. Two years after the change, guards reported that many cyclists were speeding through the park in defiance of the speed limit. When a new *Avis au public* was posted, stating that bicyclists were to limit their speed to that of a horse and carriage at a walk, many simply ignored the rule, adhering to it only when the guards were watching. The situation became increasingly dangerous, particularly in areas where considerable numbers of children were at play. A clash of park goers' interests occurred on 28 May 1890 while the Briens family from the rue du Département was visiting. The couple had brought their young son to the park to play on that warm spring day. While the youngster and his parents were walking along one of the paths, a cyclist came from behind them suddenly at a very high rate of speed, ran over the boy, and sped off. The child's clothes were torn, and he sustained bruising and several broken teeth. In his report on the matter, the inspector for the area advised that, in light of this accident, the municipal council should reconsider the 1888 statute and prohibit bicycles in the Parc des Buttes Chaumont in order to prevent something like this accident from ever happening again.[69]

The Municipal Council did, indeed, reinstate the ban on the use of bicycles by individuals in the Parc des Buttes Chaumont in late summer of 1890. Within a few weeks, the park service received a petition from a M. Richard asking if this interdiction could be modified.[70] The ban had been a hardship, he claimed, for him and for other bike owners who lived in the nearby quartier known as Combat. Richard wanted the park service to grant a kind of temporary license. He was, perhaps, aware of the licensing of fishermen in the park and may have believed that the park service would be amenable to a similar solution. According to Richard, a limited number of licenses should go to people who were "accustomed to using a bicycle and who own [rather than rent] their machines," implying that the accident(s) had been caused by novices or renters, not responsible cyclists.[71] The permissions would allow skilled riders in the park at times when there were fewer children; Richard suggested 5:00 a.m. until 8:30 a.m. But fishing and cycling were two very different activities in the eyes of the park service. In his response, Inspector Lion countered, "Contrary to what the petitioner says, there are children in all parts of the Parc des Buttes Chaumont from opening on and regardless of how few permissions might be granted, they could still give rise to accidents similar to those that occurred prior to the [reissued] ban."[72] The interests of cyclists and those of the children had clashed in Buttes Chaumont, and the Municipal Council and the Service des Promenades et Plantations sided with the children again—at least, as it concerned individual, unaffiliated cyclists and unorganized cycling. The ban did not include closely managed events such as periodic bicycle-club races. The park service had denied permission for such races in 1868, but by the 1890s, the popularity of the races made them a mainstay entertainment. The cycling clubs had become well organized, the races carried no admission for the public, and they were carefully orchestrated with monitors stationed all along the route to reduce significantly the risk of injury or inconvenience to park goers. These cycling groups learned to work with the park service to achieve their goals, and they were more successful than were individual petitioners such as M. Richard.[73]

Still, and for the time being, individual cyclists could not operate in the Park des Buttes Chaumont, and the 1896 reformulation of the city's laws on cycling in particular parks did little to encourage them. The 1896 statute further clarified how cyclists should operate in those parks, outlining details such as the right of way of carriages and pedestrians, how to operate safely in a crowd, the proper speed at intersections and corners, how to pass, what do to when confronted by a skittish horse, and so on.[74] However, the statute continued to exclude the Park des Buttes Chaumont. In response to this omission, the president of the *Touring-Club de France* championed the cause of the cycling residents of Belleville in 1897. The Touring-Club de France was the largest cycling club in France. Modeled after the English Cyclist Touring Club and a forerunner to Michelin, the group that actively promoted tourism in France through hiking, cycling, and later automobile touring. Although it professed to be apolitical, the club's political philosophy was decidedly Solidarist.[75] Since Solidarists held that acknowledgment of shared interests among individuals and social responsibility could be a viable alternative to class conflict, it is not surprising that the ostensibly middle-class touring club came to the defense of the working-class *bellevillois* at this time, and pressed the city to allow bicycles in the Parc des Buttes Chaumont, at least on the transverse road that linked rue Fessart and rue Secrétan. This was a city road, a public way, the club's president argued, and children did not play in the road. In a report on the matter, Inspector Lion indicated that he had no objections to this compromise; neither did his superior, M. Pfeiffer. However, the latter of the two added an addendum to the report reminding Lion that the recent municipal statute governing cycling in the parks had explicitly stated that individual cyclists were *only* allowed in the Bois de Boulogne, Bois de Vincennes, and Parc Monceau. Any modification to that policy adopted by the park service would require a revision by the prefect to explicitly *include* the Parc des Buttes Chaumont.[76] On 24 January 1898 the prefect, Justin de Selves, signed the amendment to the statute and the Parc des Buttes Chaumont was added to the list of parks in which open cycling would be allowed.[77] Although it is unclear whether individual riders in Belleville had solicited the assistance of the Touring-Club de France, what is evident is that, near the turn of the century, carefully organized campaigns by large sporting groups concerning the use of the parks proved much more successful than individual or smaller-group requests.

By the beginning of the twentieth century, just as Napoleon III and early city planners had hoped, Parisian greenspaces had become a significant attraction to crowds of foreign visitors, and also served the needs of citizens, the city, and the nation. The parks and squares of Paris became part of what the city offered to an increasingly expansive tourism industry. Thus, the Touring-Club de France, along with other national groups, continued to concern themselves with issues surrounding the parks, and often allied with Parisian residents to protect them. As the population of Paris grew from the start of the greenspace development program during the Second Empire to the first decade of the twentieth century, and as neighborhoods became more populous and urban dwellers possessed more leisure time, the involvement of Parisians in the details of their local greenspace likewise increased in scale and became more organized. Within the emergent mass society, groups such as skaters, sportsmen, fishermen, cyclists, and others from all over the city found in greenspaces centers for their own particular brand of communitas. In their use of the space for recreation, they experienced the spontaneous sense of unity and community that Turner outlined. This unity transcended the physical and at times the social geographies of the city, as Parisians came to identify with particular activities in the parks and squares even more, in some cases, than with individual neighborhoods. It quickly gave way, as Turner had suggested, to an organizing impulse, and these groups found ways to press their interests with the Service des Promenades et Plantations and the city administration. The concerted efforts by these entities, and on their behalf, sometimes clashed with those of other city residents who similarly sought to define the parks and squares as best befit their desires, needs, and identities. Thus, greenspaces continued to be negotiated spaces. Despite their differences, however, by the early twentieth century these competing groups came to find common ground in the defense of greenspace when ever-increasing urban development threatened it. In this way, Parisian parks and squares became the nexus of an even broader community—Parisians—with disparate interests, yet united in a desire to use and protect public greenspace, reaching beyond the quartiers and, to a certain extent, through social divisions, thus sowing the seeds of a modern preservation movement.

CONCLUSION

By the early twentieth century, approximately a half-century after the extensive greenspace development program began in Paris, parks and squares had become so integrated into city living that reducing any amount of them became unthinkable for the Parisian public. In 1912, a diverse coalition of city residents fought successfully to prevent the subdivision of the Observatoire garden, once part of the larger Luxembourg garden, and the construction of a new road and housing there. With support from neighborhood residents, doctors, mothers, sporting groups, architects, and those interested in tourism, the *Société de médecine publique et de génie sanitaire* formed a commission to protect the garden, and to secure much more. The resolution they drafted demanded that the government and the city "pledge to protect the gardens of the *Observatoire de Paris* against any subdivision."[1] The signatories further insisted that "measures be taken so that no bit of garden or open space belonging to the State or to the city of Paris, no matter how miniscule, could ever, under any pretext, be designated for construction purposes."[2] Their united approach was successful. The integration of parks and squares into city life that prompted the observatoire protest began long before the turn of the century. The roots of Parisian citizens' relationship to urban greenspace began with development of the forms themselves and evolved throughout the decades of the second half of the nineteenth century.

Adolphe Alphand's parks and squares, rigorously designed *à l'anglaise,* scientifically conceived, and technologically engineered, represented the modern marriage of the French and English early modern landscape sensibilities. The *gloire* of the French nation and the supposed egalitarianism of Bonapartism mixed with an appreciation of "natural" forms and the individual's place in the park in this new design expression. Alphand's greenspace constructions demonstrate a conspicuous awareness on the part of the engineer that this was artificial nature—hand railings and steps of *béton armé* (a new form of reinforced concrete) carved to look like wood or rock, stuccoed cliff faces, and cement-lined and -edged ponds fill the greenspace plans. He appeared as conscious of his manipulation of form as André Le Nôtre was of his manipulation of perspective in the seventeenth century. Using new, modern engineering technologies, Alphand took landscape design to another level. He went beyond the rarified realm of aesthetics and impressions and placed it in the modern, challenging world of the crowded urban environment. The park service worked to create social habitats that allowed for dramatic adjustments of spatial organization and public participation—spaces that were not just static greenspaces but that shaped and were shaped by the whole of Parisian society.

This rather synthetic form was distinguished by its functionality. It could simultaneously address the public good and some of the aims of the French state during the nineteenth century. As part of the redesign of Paris, the parks, from Napoleon III's perspective, could help to reestablish France on the world stage as a great, modern, innovative nation, with a beautiful and impressive capital. The renovation of the capital would, in the emperor's view at least, make obvious to France and to the world an end of the tribulations of revolution and the fulfillment of the Napoleonic goal of order and stability. Infused with the optimistic positivism of the day and scientific evidence of the healthful benefits of verdure in the city, these parks could also improve the quality of life for all city dwellers. At the same time, they might provide for the health and welfare of France's passive citizens (women and children), combat disease, and help secure the nation's future with a vigorous and fit youth. With Alphand at its head, the park service enjoyed a surprising degree of autonomy in its work to fulfill the promise greenspace held for all. His background and his genius as a civil engineer set him on a path to great success in his career, during which he furthered his own particular democratic vision of urban greenspaces, a vision imbued with the pragmatism and problem solving of engineering. As he rose within city government, Alphand never lost sight of his own philosophy of the justness of providing unfettered use of greenspace to the greatest number of the population with preference to none. He remained convinced of the ameliorative effects of verdure, parks, and squares, against the appalling conditions of modern urban life, and the injustice many experienced. His success in creating accessible public greenspaces earned him the respect and admiration of citizens and administrators alike. Still early in his career and in terms of the greenspace development program in Paris, Alphand understood well the significance and uniqueness of what was occurring in the capital city. Writing in 1867, he reflected on the redesign of the city and the expansion of greenspace there. Paris, he declared, had "taken the initiative in these ameliorations [planted spaces] and provided an example of the transformation of an old city, not through realizing works of fantasy and of vain pageantry, but by applying the conquests of science and of art to the viability and salubriousness of a large city."[3]

With skill, personality, and vision, Alphand also shaped the internal culture of the park service, creating an agency focused on solutions as well as fair and responsive management of the parks and squares. With newly engineered greenspace forms emerging throughout the city and open to all without the customary restrictions of the past, and an exploding, socially striated population seeking to carve out spaces of exclusivity and identity, the park service fought to maintain its (and Alphand's) first principles of greenspace as truly egalitarian. Alphand's personal conviction about the necessary publicness of urban greenspaces and his significant contribution to the nature of their development in nineteenth-century Paris has, until now, gone largely unnoticed, limiting our understanding of the renovation of Paris and the urban environment.

Beyond Paris, the work of the *Service des Promenades et Plantations* played a key role in the professionalization of landscape architecture and the creation of an international community of engineers, designers, and urban planners—a community characterized by significant transnational collaboration and sharing. Members of the park service left their mark on cities around the world, constructing parks and influencing urban planning on a global scale, in places as diverse as Phnom Penh, Cairo, and Buenos Aires. Many of the resulting projects were easily recognizable as having been created in or influenced by the Parisian style. The chief horticulturalist of the park service, Jean-Pierre Barillet-Deschamps, designed parks and gardens in Italy, Belgium, Austria, Prussia, and famously in Cairo at the request of the Khedive Ismail Pasha. Édouard André, Alphand's *jardinier principal,* designed public parks throughout continental Europe, Great Britain, and the Americas, including the Luxembourg Park in

Luxembourg City and Sefton Park in Liverpool, and parks in Rome, Monte Carlo, and Montevideo. In 1858, Adolphe Alphand was himself in communication with the New York City park commissioners and under consideration to lead their efforts to build a central park. Indeed, Frederick Law Olmsted visited Paris in 1859 and consulted with Alphand, who led him on several personally guided tours of the Bois de Boulogne. Olmsted later urged his own assistant, Charles Eliot, to travel to Europe to study the public parks there in 1885. During the trip, Eliot spent a good part of the spring and summer in Paris and conferred with Édouard André, assistant to Alphand, concerning the Boston Metropolitan Park System. Landscape architect Jules Charles Thays, who worked in the park service as André's assistant, eventually became director of parks and promenades in Buenos Aires in 1891. Jean-Claude Nicolas Forestier, a landscape architect who trained under Alphand and continued his policies as chief conservator after the director's death in 1891, designed projects in Cuba, Spain, and Morocco. Forestier, Alphand, and André all wrote important and well-known treatises on the history, theory, and practice of landscape design. Paris and the park service thus played a significant role in the development of urban planning and landscape architecture well beyond Paris. This suggests new avenues of research and raises new questions about the ways in which the French model of design and use of greenspaces was "translated" into different patterns of meaning and use by other city governments and city residents.

The expansion of Parisian public greenspace during the nineteenth century meant that administering those spaces was quite naturally uncharted territory for the municipal government, as increased accessibility to public greenspaces was something quite new for Parisians. Together, the park service and city residents, park builders and users, participated in working out the place that those parks and squares would hold in peoples' lives, and, in a remarkable form of community engagement, transformed a linear design-toward-use model into a reciprocal, interactive, dynamic, and fluid process. Thus, the distinctions between park creator and park user blurred, and greenspace became a negotiated space in which the park goer could implement material or experiential changes either through a direct challenge or through spatial practices. Michel de Certeau's articulation of how spatial practices can imbue a place, or a *lieu,* with meaning, transforming it into a space, or an *espace,* is then particularly applicable here.[4] For de Certeau, however, the force that individuals exert on a space to alter its nature and meaning, the spatial practice, is secretive, subversive, private, and unmappable. Certainly some spatial practices remain forever invisible to historians. Yet the engagement of ordinary Parisians, their movement through space or thought, was quite perceptible—often collective. Indeed, the park goer's role in the development of relevant, urban, public greenspace in Paris was both energetic and fruitful.

There is much to learn about the park users and greenspaces from a study of the implementation and evolution of the park building program that factors in how the spaces were received and used. A close examination of public greenspaces and the lived experience of city residents reveals that, although Parisians may have had very little direct control over planning decisions, they were not passive onlookers as their cityscape underwent tremendous change during the nineteenth century. On the contrary, piecing together the relationship between the city and the park goers throughout the Second Empire and into the Third Republic, this research has uncovered the very active role ordinary citizens played in shaping those new spaces. At times, this engagement began during construction and lasted long after its completion. Residents living near a park or square pressed authorities to alter the design or the operation of these spaces, making them more relevant in their lives and weaving them into the fabric of family and community life. Indeed, if democratization of greenspace occurred, it owed much to the way in which average Parisians of various social classes demanded it. A multiplicity of urban communities came together, sometimes spontaneously, in and around the park space, and they grew increasingly organized and persistent in their efforts to make known their requirements. The process was not seamless or without contestation, but it was participatory. It is this public participation that was a crucial component of greenspace development then, and remains vital today as municipalities work to engineer greenspaces that address the fullest array of public use and necessity.

Through its extensive park development program focused on serving the needs of city residents and ameliorating the conditions of modern urban living, the work of the park service represents a link between the earlier isolated projects of the Reform movement in Great Britain in the 1840s (Joseph Paxton's Birkenhead Park in Liverpool, for example) and the late-nineteenth, early twentieth-century planning initiatives, which met with varying degrees of success, such as the City Beautiful Movement in the United States and the modernist Siedlungen in Germany. Although the motivation and ethos of these planning efforts may have differed, communal greenspace and quality of life were always essential components.

Twentieth-century conceptions of the relationships among nature, the built environment, and quality of life continued to evolve as the pace of social, economic, and technological change accelerated, and as new questions arose concerning greenspace and society. In 1952, against a backdrop of increasing post–World War II disintegration of urban centers, the Congrès Internationaux d'Architecture Moderne (CIAM) offered a concept of a new, modern city—one in which greenspace could function to fully integrate the social, economic, and cultural spaces of the modern city, as well as link the city to the surrounding region. Greenspaces would not constitute islands of verdure, as they appeared Paris, but rather the essential fabric of the city serving in a new way to bind the whole together—the physical and experiential.[5] Realization of that vision, however, proved harder to achieve. Rather than planned integration, environmental movements of the 1960s placed nature and humanity in opposition to preserve "nature" and "wilderness" (defined in vastly different ways) from the threats posed by economic and technological growth. The natural world then had to be protected from the intrusive hand of humanity. Yet, as historian Michael Bess argued, the very efforts to maintain the delineation between man and nature, and the success of the environmentalists' message in the form of increased intervention and stewardship in France and elsewhere, resulted in a blurring of that boundary, the emergence of a spectrum of "natural-artificial hybrids," and new ways of thinking about society and nature.[6]

In recent years, municipalities have paid a great deal of attention to issues surrounding this complex relationship and understanding of nature, particularly in the urban milieu. Discussions about quality of life in cities, public health and welfare, community formation, effective public outreach, and greenspace expansion abound in light of the continued increase in urban populations and density. There are countless park building programs underway in the second decade of the new millennium that reflect understandings of nature shaped by twentieth-century movements, and echo concerns and developments in nineteenth-century Paris. Some examples are New York City's *PlaNYC 2030*, dedicated to providing public greenspace within a ten-minute walk for all city residents; the many *Million Trees* initiatives in cities throughout the United States seeking to increase urban forests in major metropolitan areas; the urban revitalization program *Urban Releaf* in Oakland, California, which is committed to revitalizing blighted neighborhood spaces and reinvigorating communities through citizens' participation in planting and maintaining greenspace; and the concern over greenspace loss to urban development in international cities like Istanbul and Shanghai. All of these programs seek to alter the built environment and, crucially, engage the public's participation in shaping or reshaping urban space through greenspace development.

The image that then emerges of public greenspace in Paris during the nineteenth century through an examination of the relationships among the park goers, workers, advocates, planners, and administrators is something quite different from Georges Seurat's rather stiff, ossified view of an afternoon on the Île de la Grande Jatte. In the physical and experiential aspects of Parisian greenspace, one can see all of the diversity and dynamism that characterized the French capital city during a time of tremendous urban growth and economic, social, and political change. Every stratum of society strove to define and inhabit the parks, from elites to those suffering on the margins of society, and each won something and lost something. In this impressive mélange there were reformers, doctors, children, *mères de famille*, irreverent adolescents, working women, guards, business owners, elites, skaters, prostitutes, administrators, cyclists, workers, criminals, fishermen, the homeless, and the indigent. All approached the space from their own unique perspective and in ways large and small engaged in the process of engineering this malleable, "natural" space and furthering (for some) a decidedly democratic understanding of it. In doing so, together they established firmly a place for constructed nature in the urban environment of Paris and in the lives of Parisians—a place that continues to have implications for city residents today, and one that can inform urban planning initiatives well beyond the French capital.

APPENDIX
PARKS AND SQUARES

Arrondissement	Greenspace	Inaugurated or Redesigned
1	Square Henri IV	1884
	Square du Carrousel	1883
	Square des Innocents	1859
2	Square Louvois	1859
	Préau de la Bourse	1857
3	Square du Temple	1857–59
	Square des Arts et Métiers	1858
4	Square de la Tour St. Jacques	1856
	Jardin de l'Hôtel de Ville	1865
5	Place Walhubert	1863
	Fontaine Soufflot	1864
	Square des Arènes	1892
	Square St. Médard	1875
	Square Monge	1868
6	Square de l'Observatoire	1867
	Jardin du Luxembourg (including Jardins de l'avenue de l'Observatoire)	1865–67
7	Parc du Champ de Mars	1878/1889
	Square Boucicaut	1870
	Square des Ménages	1865
	Square Ste. Clothilde	1857–59
	Square des Invalides	1865
	Square de la Tour Maubourg	1865

Arrondissement	Greenspace	Inaugurated or Redesigned
8	Jardins des Champs-Élysées	1858–61
	Parc Monceau	1861–63
	Square Delaborde	1867
	Square Louis XVI	1867
	Place de l'Europe	1866
9	Square de la Trinité	1865–67
	Square Montholon	1863–64
	Square d'Anvers	1877
	Square Vintimille	1861
10	Jardinets du Canal St. Martin	1891
	Square St. Vincent de Paul	1867
	Square St. Laurent	1896
	Square du Combat	*
11	Square Parmentier	1872
	Place du Prince Eugène	1864
	Jardinets du boulevard Richard Lenoir	1861–63
12	Square de la Mairie	1877
	Square Jean Morin	1900
	Square Carnot	1889
	Square de l'hôpital Trousseau	1880
13	Square de la Butte aux Cailles	*
	Place d'Italie	1860–78
	Jardin de la statue de Pinel	1879
14	Parc Montsouris	1875
	Square de Montrouge	1862
	Squares de Denfert-Rochereau (3)	1880–96
15	Square Violet	1860
	Square de la place du Commerce	1873
	Square Victor	1865–67
	Square Cambronne	1881
	Square Garibaldi	1881
	Square de Vaugirard	1861–62
	Place de la Mairie de Grenelle	1862–63
	Square Adolphe Chérioux	1896

Appendix: Parks and Squares

Arrondissement	Greenspace	Inaugurated or Redesigned
16	Jardin du Ranelagh	1860
	Jardins de l'avenue de l'Impératrice	1855
	Square du Marché de Passy	1867
	Square Lamartine	1862
	Parc du Trocadéro	1878
17	Square des Batignolles	1862
	Square Malesherbes	1862
	Square des Épinettes	1893
18	Square Jessaint	1876
	Square de la Chapelle	1862–63
	Square Montmartre	**
	Square St. Pierre	1892
	Square St. Bernard	1891
19	Square de Belleville (Pl. des Fêtes)	1862
	Square de l'hôpital Tenon	1879
	Parc des Buttes Chaumont	1863–67
	Esplanade des abattoirs de la Villette	1867–68
20	Square de la Mairie	1879
	Square de l'avenue Gambetta	1889
	Square de la place de la Réunion	1862
Les bois	Bois de Boulogne	1852–58
	Bois de Vincennes	1858–68

Sources: AP, TriBriand VM 19, Ville de Paris, Service des Promenades et Plantations, *Tableau statistique des bois, parcs, squares, promenades publiques plantées, 28 août 1868*; Paul Joanne, ed., *Dictionnaire géographique et administratif de la France* (Paris: Librarie Hachette et cie., 1898), 3189; Mairie de Paris-Parcs, Jardins, et Squares, www.paris.fr/pratique/paris-au-vert/parcs-jardins-squares/p4952.

* Appears in 1898 Joanne inventory (no date).
** Appears in 1898 Joanne inventory with notation "*non terminé.*"

NOTES

PREFACE

1. Paul Imbs and B. Quemada, *Trésor de la langue française; dictionnaire de la langue du XIXe et du XXe siècle, 1789–1960* (Paris: Editions du Centre national de la recherche scientifique, 1971); Édouard André, *L'art des jardins: Traité général de la composition des parcs et jardins* (1879; rpt. Marseille: Lafitte reprints, 1983).

INTRODUCTION

Epigraph: William Robinson, *The Parks, Promenades, and Gardens of Paris Described and Considered in Relation to the Wants of Our Own Cities and of Public and Private Gardens* (London: J. Murray, 1869), 90.

1. David Pinkney, *Napoleon III and The Rebuilding of Paris* (Princeton, N.J., Princeton University Press, 1958).
2. See for example David Jordan, *Transforming Paris: The Life and Labors of Baron Haussmann* (New York: Free Press, 1995); and Michel Carmona, *Haussmann* (Paris: Fayard, 2000).
3. Emblematic of this approach: David Harvey, *Paris, Capital of Modernity* (New York: Routledge, 2003); T. J. Clark, *The Painting of Modern Life: Paris in the Art of Manet and His Followers* (Princeton, N.J.: Princeton University Press, 1984); and Nicholas Green, *The Spectacle of Nature: Landscape and Bourgeois Culture in Nineteenth-Century France* (Manchester, U.K.: Manchester University Press, 1990).
4. See John Dixon Hunt, *The Figure in the Landscape: Poetry, Painting, and Gardening during the Eighteenth Century* (Baltimore: Johns Hopkins University Press, 1976); *Gardens and the Picturesque: Studies in the History of Landscape Architecture* (Boston: MIT Press, 1992); and *Tradition and Innovation in French Garden Art: Chapters of a New History* (Philadelphia: University of Pennsylvania Press, 2002), co-edited with Michel Conan.
5. Yi-Fu Tuan, "Thought and Landscape," in *The Interpretation of Ordinary Landscapes: Geographical Essays,* ed. D. W. Meinig and John Brinckerhoff Jackson (Oxford, U.K.: Oxford University Press, 1979).

6. The absence of police records for the earlier decades of this study owing to the fire at the Hôtel de Ville in 1871, and missing registers for the individual commissariats in the *quartiers* (CB series at the Archives de la Préfecture de Police de Paris) make the *Service des Promenades et Plantations* guard reports essential to understanding a range of proscribed activity in the parks.

7. Nicholas Papayanis, *Planning Paris before Haussmann* (Baltimore: Johns Hopkins University Press, 2004).

8. See Mary Pickering, *Auguste Comte: An Intellectual Biography* (Cambridge, U.K.: Cambridge University Press, 1993); W. M. Simon, *European Positivism in the Nineteenth Century: An Essay in Intellectual History* (Ithaca, N.Y.: Cornell University Press, 1963); D. G. Charlton, *Positivist Thought in France During the Second Empire, 1852–1870* (Oxford, U.K.: Clarendon Press, 1959); and Émile Littré, *Auguste Comte et la philosophie positive* (Paris: L. Hachette et cie., 1864), 662–81.

9. See Auguste Comte and Gertrud Lenzer, *Auguste Comte and Positivism: The Essential Writings* (New York: Harper & Row, 1975), xxiv–lv; and Auguste Comte and Frederick Ferré, *Introduction to Positive Philosophy* (Indianapolis: Bobbs-Merrill, 1970), 2–4.

10. Michel de Certeau, *The Practice of Everyday Life* (Berkeley: University of California Press, 1984), 93; Ian Buchanan, *Michel de Certeau: Cultural Theorist* (London: SAGE Publishing, 2000), 114.

11. De Certeau, *The Practice of Everyday Life*, 117–18.

12. Victor Turner, *The Ritual Process: Structure and Anti-Structure* (Chicago: Aldine Pub. Co., 1969), 132.

13. Ibid., 96.

14. Ibid., 138–39, 177–78.

15. Hunt, *The Figure in the Landscape*, 63. Hunt asserted that the *jardin anglais* expressed in landscape form Britain's freedom from tyranny, oppression, and autocracy while, for Britons, early modern French gardens symbolized absolutism and despotism.

16. Édouard André, *L'art des jardins: Traité général de la composition des parcs et jardins* (1879; rpt. Marseille: Lafitte reprints, 1983), 83; Joseph Disponzio, "Jean-Marie Morel and the Invention of Landscape Architecture," in *Tradition and Innovation in French Garden Art*, ed. Hunt and Conan, 135. Jean-Jacques Rousseau's rhapsodic description of an idyllic garden in *Julie, ou La Nouvelle Heloïse* inspired landscape gardener Jean-Marie Morel and his patron, the marquis de Girardin, to see it manifested it in physical form at Ermenonville, which had much to do with popularizing the *jardin anglais* style among some of the French nobility and wealthy estate owners. See Jean-Jacques Rousseau, *The Collected Writings of Rousseau* 6: *Julie, or, The New Heloise: Letters of Two Lovers Who Live in a Small Town at the Foot of the Alps*, trans. Philip Stewart and Jean Vaché (Hanover, N.H.: Dartmouth College, 1997), 387–401.

17. Dana Arnold, *Rural Urbanism: London Landscapes in the Early Nineteenth Century* (Manchester, U.K.: Manchester University Press, 2005), 52–53, 180–81; Norman T. Newton, *Design on the Land: The Development of Landscape Architecture* (Cambridge Mass.: Harvard University Press, 1971), 223–225.

18. Gilles-Antoine Langlois, *Folies, tivolis et attractions: Les premiers parcs loisirs parisiens* (Paris: Délégation à l'action artistique de la ville de Paris, 1991), 15–21; Louis Jacob Frazee, *The Medical Student in Europe* (Maysville, Ky.: R. H. Collins, 1849), 42; and Adolphe Alphand and Alfred Auguste Ernouf, *L'art des jardins. Parcs—jardins—promenades—étude historique—principles de la composition*

des jardins—plantations—décoration pittoresque et artistique des parcs et jardins publics; traité pratique et didactique (Paris: J. Rothschild, 1886), 297–304.

19. Archives Nationales (hereafter AN) F14 11459, Letter from Andrew H. Green, President of the New York City Board of Park Commissioners, to M. Adolphe Alphand, 23 December 1858.

20. Archives de Paris (hereafter AP) Perotin 10653 54, *Oiseaux aquatiques,* Rapport du Conservateur, Ponts et Chaussées, Département de la Seine, Ville de Paris, Service Municipal des Travaux Publics de Paris, Service des Promenades et Plantations, 5 December 1860.

21. Hunt, *Gardens and the Picturesque,* 13.

CHAPTER ONE

Epigraph: Jean-Gilbert-Victor Fialin Persigny and Henri de Laire, *Mémoires du duc de Persigny* (Paris: E. Plon, Nourrit & cie., 1896), 256.

1. Napoleon III, quoted in Michel Carmona, *Haussmann* (Paris: Fayard, 2000), 16.
2. William H. C. Smith, *Napoléon III: The Pursuit of Prestige* (London: Collins and Brown, 1991), 59–61; David Harvey, *Paris, Capital of Modernity* (New York: Routledge, 2003), 112–13; Pierre Pinon, *Paris, biographie d'une capitale* (Paris: Hazan, 1999), 190; and Colin Jones, *Paris: Biography of a City* (New York: Viking, 2005), 300–301.
3. David Jordan, *Transforming Paris: The Life and Labors of Baron Haussmann* (New York: Free Press, 1995), 166.
4. Jean-Jacques Berger, quoted in Jordan, *Transforming Paris,* 166.
5. Carmona, *Haussmann,* 221–20.
6. Jordan, *Transforming Paris,* 166–67.
7. Carmona, *Haussmann,* 21.
8. During the French Revolution, the National Convention had eliminated the Paris Commune and with it the office of mayor in 1794 to reassert the authority of the national government over what it deemed radical elements in the city and to prevent future challenges. The government of the early Second Republic reestablished the office briefly in 1848 only to do away with it again that same year. Similarly, after the fall of the Second Empire, once again the national government reestablished the office in 1870 only to abolish it again in 1871. The city of Paris then functioned without a mayor from 1871 until the office was reestablished in 1977 and the election of Jacques Chirac.
9. Jordan, *Transforming Paris,* 176–78.
10. David Pinkney, *Napoleon III and the Rebuilding of Paris* (Princeton, N.J.: Princeton University Press, 1958), 44.
11. "Square du Temple," *Le Monde illustré* 30 (7 November 1857): 11 (emphasis added).
12. Ibid.
13. Ibid.
14. Germaine Boué, *Les squares de Paris: La tour Saint Jacques* (Paris: Librairie centrale, 1864), 6.
15. Ibid., 7. A *quartier* is an administrative municipal division approximating one-quarter of an arrondissement, although the term is also often used to denote the social community in Parisian neighborhoods—a kind of urban village.

16. Ibid., 12.

17. Alexandre Jouanet, *Mémoire sur les plantations de Paris* (Paris: Imprimerie Horticole de J.-B. Gros, 1855), 18.

18. Alexandre Jouanet, *Paris et ses plantations* (Paris: Typographie de Morris et cie., 1860), 7.

19. Ibid., 8.

20. See for example, Patrice de Moncan and Claude Heurteux, *Le Paris d'Haussmann* (Paris: Mécène, 2002); also Carmona's *Haussmann* and Jordan's *Transforming Paris*.

21. Chantal Auré, *Paris XIXe–XXe siècles, Urbanisme, architecture, espaces verts: Bibliographie et sources imprimés à la Bibliothèque des Archives de Paris* (Département de Paris: Archives de Paris, 1995), 495–496.

22. Pinkney, *Napoleon III and the Rebuilding of Paris*, 47.

23. Georges Lafenestre, *Notice sur la vie et les oeuvres de M. Alphand par M. Georges Lafenestre, membre de l'Académie des Beaux Arts, lue dans la séance du 29 juillet 1899* (Paris: Typographie de Firmin-Didot et cie., 1899), 6.

24. AN, F8 11459, *Chronique du service, Alphand, Jean-Charles Adolphe*, Ministère de Travaux Publics, Département de la Seine, Service municipal de Paris, last entry 1891.

25. AN, F8 11459, *Dossier personnel de M. Alphand*, Ministère de l'Agriculture, du Commerce, et des Travaux Publics, Département de la Seine, Service municipal des Travaux Publics de Paris, 1868.

26. Georges-Eugène Haussmann, *Mémoires du Baron Haussmann 1: Grand Travaux de Paris* (Paris: Guy Durier, 1979), 126.

27. Ibid.

28. Lafenestre, *Notice sur la vie et les oeuvres de M. Alphand*, 6.

29. Ibid.

30. Ibid., 10.

31. Adolphe Alphand, "Discours de M. Alphand sur M. le Baron Haussmann" in Haussmann, *Mémoires* 1: xv.

32. Ibid., 131.

33. Ibid., 183.

34. Adolphe Alphand, *Les promenades de Paris: Histoire—description des embellissements—dépenses de création et d'entretien des Bois de Boulogne et de Vincennes, Champs-Élysées—Parcs—squares—boulevards—places plantées, Étude sur l'art des jardins et arboretum, 487 gravures sur bois, 80 sur acier, 23 chromolithographies* (1867; rpt. Princeton, N.J.: Princeton Architectural Press, 1984), lviii.

35. Ibid., 126.

36. Ibid., 126–27; and Adolphe Joanne, *Paris illustré en 1870 et 1876; guide de l'étranger et du Parisien contenant 442 vignettes dessinées sur bois, un plan de Paris et quartorze autres plans et un appendice pour 1876, Troisième édition* (Paris: Hachette, 1876), 243–44.

37. Édouard André, "Les jardins de Paris" in *Paris guide par les principaux écrivains et artistes de la France, Deuxième partie: La vie* (Paris: A. Lacroix, Verboeckhoven et cie., 1867), 1215.

38. Ibid., 130.

39. Alphonse du Breuil, *Cours élémentaire théorique et pratique d'arboriculture, comprenant l'étude des pépinières d'arbres et d'arbrisseaux forestiers, fruitiers et d'ornement; celle des plantations d'alignement forestières et d'ornement; la culture spéciale des arbres à fruits, à cidre et de ceux à fruits de table* (Paris: Langlois et Leclercq, 1853), frontispiece.

40. Joseph Paxton and John Lindley, *A Pocket Botanical Dictionary; Comprising the Names, History, and Culture of All Plants Known in Britain* (London: Bradbury and Evans, 1853), s.v. "platanus"; D. J. Mabberley, *The Plant-Book: A Portable Dictionary of the Higher Plants Utilising Cronquist's* An Integrated System of Classification of Flowering Plants (*1981*) *and Current Botanical Literature, Arranged Largely on the Principles of Editions 1–6 (1896/97–1931) of Willis's* A Dictionary of the Flowering Plants and Ferns (Cambridge, U.K.: Cambridge University Press, 1987), s.v. "platanus."

41. Jouanet, *Paris et ses plantations*, 10.

42. "Le Square Montholon," *Le Monde illustré* 14, no. 353 (16 January 1864): 48.

43. AP, 1304W 162, *Square du Temple, Complément de plantations, Arbres à prendre dans différent cimetières*, Rapport du conducteur principal inspecteur, Ponts et Chaussées, Département de la Seine, Ville de Paris, Direction des Travaux de Paris, Service des Promenades, 28 October 1875.

44. De Moncan and Heurteux, *Le Paris d'Haussmann*, 209.

45. David C. Goodman and Colin Chant, *European Cities & Technology: Industrial to Post-Industrial City* (London: Routledge, 1999), 112.

46. Alphand, *Les promenades de Paris*, 52.

47. Ibid. The expansive lawns in the park at the Chateau de la Muette were an anomaly because they flourished without significant effort on the part of the city thanks to the deer that grazed there, fertilizing the soil and building the topsoil.

48. Ibid., lviii.

49. Ibid., 52 (emphasis added).

50. Ibid.

51. Ibid., 54. The maintenance of grass in Parisian parks and squares remains a challenge today, and visitors to the French capital quickly become familiar with the ubiquitous *Pelouse au repos* signs, asking park goers politely to "Keep off the grass."

52. AP, Perotin 10653 90, *Copie d'une circulaire plus spéciale aux bois de Boulogne et Vincennes*, Ponts et Chaussées, Département de la Seine, Ville de Paris, Direction des Travaux de Paris, Promenades, Éclairage, Voitures et Concessions, 24 October 1874.

53. AP, Perotin 10653 86, *Entretien des jardins et squares, Établissement d'une gazonnière dans les fortifications Porte Maillot,* Rapport du l'ingénieur ordinaire, République Française, Ponts et Chaussées, Direction des Travaux de Paris, Service des Promenades, 4 January 1884.

54. AP, Perotin 10653 54, Rapport du jardinier en chef, Ponts et Chaussées, Ville de Paris, Service spécial des Promenades et Plantations, 5 April 1857.

55. Ibid.

56. AP, Perotin 10653 54, *Oiseaux aquatiques,* Rapport du conservateur, Ponts et Chaussées, Département de la Seine, Ville de Paris, Service municipal des Travaux Publics de Paris, Service des Promenades et Plantations, 5 December 1860.

57. Ibid.

58. AP, 10653 67, *Bois de Boulogne, Vente de poissons,* Rapport du conservateur du Bois de Boulogne, Ponts et Chaussées, Département de la Seine, Ville de Paris, Direction de la Voie publique et des Promenades, Service des Promenades et Plantations, 9 September 1871. This requisitioning led to a heated battle which lasted over a year concerning who in government held ultimate authority over park resources. The conflict involved the Service des Promenades et Plantations, the city, the Ministry of Agriculture, and the military.

59. AP, 1304W 113 *Bois de Boulogne et de Vincennes, Pêche à la linge,* Extrait du registre des procès verbaux des séances du Conseil municipal de la Ville de Paris, Préfecture de la Département de la Seine, Direction des Travaux de Paris, Séance du 13 mars 1872; AP, 1304W 113, *Parc des Buttes Chaumont, Demande d'un permis de pêche, Sieur Pierre Charles Dubois,* Rapport du conducteur principal, Service technique des Promenades, 21 June 1876.

60. AP, 2380 20, *Parc de Montsouris; Pétition pour la délivrance de permis de pêche à la ligne dans le lac,* Rapport du jardinier du conducteur, République Française, Préfecture de la Seine, Direction administrative des Travaux de Paris, Voirie et Travaux d'Ingénieurs, Service technique des Promenades, 9 April 1897.

61. Alphand, *Les promenades de Paris,* 110–13.

62. *Le Monde illustré* 2, no. 46 (27 February 1858); see also AP, TriBriand 118, *Parc de Montsouris; Enlèvement de glace pendant l'hiver 1896–1897, Demande de l'exonération de paiement de redevance,* Rapport de Inspecteur [sic], République Française, Préfecture de la Seine, Direction administrative des Travaux de Paris, Voirie et Travaux d'Ingénieurs, Service technique des Promenades, Inspection des Promenades, 27 February 1897.

63. "Glacières," *L'Illustration journal universel* 31 (January–June 1858): 237–38.

64. AP, VONC 291, Letter from Ch. d'Alleziette, Inspecteur des Plantations to M. l'Inspecteur en Chef, 10 November 1865.

65. Ibid.

66. AP, VONC 291, Letter from Le Jardinier en chef signée Barillet to M. l'Ingénier en chef, 10 November 1865.

67. Ibid.

68. Édouard André, *L'art des jardins, Conférence faite le 26 août 1866 à Troyes dans la grande salle de l'Hôtel de Ville sous les auspices de la Société horticole, vigneronne, et forestière* (Troyes: Imprimerie et lithographie Dufour-Bouquot, 1866), 22.

69. Ibid., 23.

70. Ibid.

71. "Le jardin Solferino à Rouen," *Le Monde illustré* 18, no. 462 (17 February 1866): 112.

72. Ibid.

73. "Les Plantations Urbaines," *La Nature* 59 (18 July 1874): 103. Also see appendix, above.

74. Archives Départmentales de la Gironde (hereafter ADG) 2 O 352, *Extrait du registre des délibérations du Conseil municipal de la Ville de Bordeaux, Séance du 18 juillet 1882,* Mairie de la Ville de Bordeaux.

75. Ibid. During the Third Republic, municipal counselors were elected by means of universal male suffrage; thus, some of the counselors in Bordeaux represented largely working-class quartiers and could be held accountable by their constituents.

76. ADG, 2 O 352, *Extrait du registre des délibérations du Conseil municipal de la Ville de Bordeaux, Séance du 4 mars 1887,* Mairie de la Ville de Bordeaux.

77. Ibid.

78. Ibid.

79. Ibid.

80. Victor Fournel, "Les Monuments du nouveau Paris," *Le Correspondant* 61 (April 1864): 878–79.

81. Ibid.

82. Fournel was referring to a plan, made public in December of 1865, which called for the extension of the rue Férou south to rue Carnot. (The rue Carnot no longer exists, but it intersected the rue d'Assas midway between the present-day rue Auguste Comte and rue Michelet on the southern end of the Jardin du Luxembourg.) Public outrage over the reduction in the size of the garden resulted in a rare intercession on the part of the emperor, who altered the plans by decree in 1866 enough to protect some of the area. Nevertheless, the Luxembourg Garden was, indeed, reduced in size. See Richard S. Hopkins, "*Sauvons le Luxembourg*: Urban Greenspace as Private Domain and Public Battleground, 1865–1867," *Journal of Urban History* 37, no. 1 (January 2011): 43–58.

83. Fournel, "Les Monuments du nouveau Paris," 879.

84. Ibid., 880.

85. Ibid., 881.

86. Ibid., 882 (emphasis in original).

87. Victor Fournel, "Voyage à travers l'Exposition Universelle," *Le Correspondant,* April 1864, 964.

88. "Le parc des Buttes Chaumont," *Le Monde illustré* 20, no. 525 (4 May 1867): 271.

89. Ibid.

90. "Exposition Internationale à Londres," *La Nature* 57 (4 July 1874): 70.

91. Maxime du Camp, *Paris ses organes, ses fonctions et sa vie dans la seconde moitié du XIXe siècle,* 3rd ed. (Paris: Librairie Hachette et cie., 1875), vol. 6: 252.

92. Eugène Pelletan, "Histoire de Paris," in *Paris guide par les principaux écrivains et artistes de la France, Prémier partie: La science, L'art* (Paris: A. Lacroix, Verboeckhoven et cie., 1867), 46–48.

CHAPTER TWO

Epigraph: Jean-Baptiste Fonssagrives, *Hygiène et assainissement des villes* (Paris: J.-B. Baillière, 1874), 175.

1. Nicholas Papayanis, *Planning Paris before Haussmann* (Baltimore: Johns Hopkins University Press, 2004), 130; Christopher Bryant, *Positivism in Social Theory and Research* (New York: St. Martin's Press, 1985) 23–26; and William Coleman, *Death Is a Social Disease: Public Health and Political Economy in Early Industrial France* (Madison: University of Wisconsin Press, 1982), 281.

2. See Catherine Kudlick, *Cholera in Post-Revolutionary Paris: A Cultural History* (Berkeley: University of California Press, 1996), chap. 2.

3. Stéphane Flachat, *Religion Saint-Simonienne, Le Choléra—Assainissement de Paris* (Paris: Everat Impr., 1832), 3. See also Papayanis, *Planning Paris before Haussmann,* 149. Papayanis provides an excellent, detailed analysis of this and other responses to the 1832 cholera epidemic that addressed issues beyond those of increased air and light and public spaces.

4. Flachat, *Religion Saint-Simonienne, Le Choléra,* 3. See also Napoleon III's "Extinction of Pauperism" in *The Political and Historical Works of Louis Napoleon Bonaparte, President of the French Republic. Now First Collected with an Original Memoir of His Life, Brought Down to the Promulgation of the Constitution of 1852; and Occasional Notes* (New York: H. Fertig, 1972), vol. 2, for a similar proposal to move work sources outside of the city and improve crowded conditions, the presumption being that workers geographically relocate to follow employment.

5. See Michel Lévy, *Traité d'hygiène publique et privée* (Paris: J.-B. Baillière, 1844); Maxime Vernois, *Traité pratique d'hygiène industrielle et administrative: Comprenant l'étude des éstablissements insalubres, dangereux et incommodes* (Paris: Chez J. B. Baillière, 1860); and Claude Lachaise, *Topographie médicale de Paris; ou, Examen général des causes qui peuvent avoir une influence marquée sur la santé des habitans* [sic] *de cette ville, le caractère de leurs maladies, et le choix des précautions hygiéniques qui leur sont applicables* (Paris: Baillière, 1822), for discussions of health and air circulation.

6. Seine (France), and Benoiston de Châteauneuf, *Rapport sur la marche et les effets du choléra-morbus dans Paris et les communes rurales du département de la Seine par la commission nommée, avec l'approbation de M. le ministre du commerce et des travaux publics, par MM. les préfets de la Seine et de police, Année 1832* (Paris: Imprimerie royale, 1834), 191 (emphasis added).

7. Louis-René Villermé, *De la mortalité dan les divers quartiers de la Ville de Paris* (n.p., 1830); Alexandre-Jean-Baptiste Parent-Duchâtelet, *De la prostitution dans la ville de Paris, considérée sous le rapport de l'hygiène publique, de la morale et de l'administration . . . précédé d'une notice historique sur la vie et les ouvrages de l'auteur, par Fr. Leuret* (Paris: J.-B. Baillière, 1836).

8. For a discussion of the nature and diversity of urban reform philosophies in the early 1800s, see Papayanis, *Planning Paris before Haussmann*, 82–95; and Rachel Fuchs, *Abandoned Children: Foundlings and Child Welfare in Nineteenth-Century France* (Albany: State University of New York Press, 1984), chap. 2.

9. See Edwin Chadwick, *Report to Her Majesty's Principal Secretary of State for the Home Office from the Poor Law Commissioners, On an Inquiry into the Sanitary Condition of the Labouring Population of Great Britain* (London: W. Clowes and Sons, 1842); and Henri-Antoine Frégier, *Des classes dangereuses de la population dans les grandes villes et des moyens de les rendre meilleures* (Paris: J.-B. Baillière, 1840).

10. Frégier, *Des classes dangereuses de la population dans les grandes villes*, 126.

11. Ibid., 127.

12. Ibid., 139.

13. Middle-class reformers throughout the nineteenth century commonly associated the cohabitation of members of the opposite sex outside of marriage with sexual promiscuity. This included siblings who slept in the same room, or worse in their eyes, in the same bed. This kind of moralizing on the part of many reformers when it came to the survival strategies that poor families used to address pure economic necessity revealed a lack of full understanding of the realities of life for the indigent and working poor, and an inherent paternalism. This disconnect often impeded the success of even the most genuine efforts. See Joan Wallach Scott, *Gender and the Politics of History* (New York: Columbia University Press, 1988); Rachel Fuchs, *Poor and Pregnant in Paris: Strategies for Survival in the Nineteenth Century* (New Brunswick, N.J.: Rutgers University Press, 1992), 39–41.

14. Frégier, *Des classes dangereuses de la population dans les grandes villes*, 143.

15. Michel Lévy, *Traité d'hygiène publique et privée* (Paris: J.-B. Baillière, 1844), 547.

16. Ibid., 549.

17. Ibid., 555.

18. Ibid., 550.

19. Ibid., 588.

20. Ibid., 591.

21. Ibid., 596.

22. On positivism and mid-century republicanism see Sudhir Hazareesingh, *Intellectual Founders of the Republic: Five Studies in Nineteenth-Century French Republican Political Thought* (Oxford, U.K.: Oxford University Press, 2001); W. M. Simon, *European Positivism in the Nineteenth Century: An Essay in Intellectual History* (Ithaca, N.Y.: Cornell University Press, 1963); and Roland N. Stromberg, *European Intellectual History Since 1789* (Englewood Cliffs, N.J.: Prentice-Hall, 1990), 109.

23. Antoine Ambroise Tardieu, *Dictionnaire d'hygiène publique et de salubrité, ou, Répertoire de toutes les questions relatives à la santé publique, considérées dans leurs rapports avec les subsistances, les épidémies, les professions, les établissements et institutions d'hygiène et de salubrité, complété par le texte des lois, décrets, arrêtés, ordonnances et instructions qui s'y rattachent* (Paris: J.-B. Baillière, 1852), 113.

24. Ibid., 299.

25. Frégier, *Des classes dangereuses de la population dans les grandes villes*, 136–37.

26. Ibid.

27. Tardieu, *Dictionnaire d'hygiène publique et de salubrité*, 113–14.

28. AN, F8 210, Letter from M. l'Administrateur général de la Bibliothèque nationale to M. le Ministre de l'Agriculture et Commerce thanking him for sending a copy of Roberts's book "publié par order de la président de la république," 8 July 1851. This carton also contains a dossier of letters from city and town officials from all over France acknowledging the receipt of their copy of the book and similarly thanking the Ministry of Agriculture and Commerce for having sent it.

29. Henry Roberts, *The Dwellings of the Labouring Classes, Their Arrangement and Construction; Illustrated by a Reference to the Model Houses of the Society for Improving the Condition of the Labouring Classes, with Other Buildings Recently Erected: and an Appendix, Containing H.R.H. Prince Albert's Exhibition Model Houses, Hyde Park, 1851; the Model Cottages etc. Built by the Windsor Royal Society; with Plans and Elevations of Dwellings Adapted to Towns, As Well As to Agricultural and Manufacturing Districts* (London: Seeleys, 1853), 57.

30. Roberts, *The Dwellings of the Labouring Classes*, 23–24.

31. Alphonse Balthazard Bertelé, *Quelques mots sur les logements des ouvriers* (Strasbourg: Impr. d'Ad. Christophe, 1863), 22.

32. See Anne-Louise Shapiro, "Housing Reform in Paris: Social Space and Social Control," *French Historical Studies* 12, no. 4 (Autumn 1982): 486–507. Napoleon III's interest in the issue and Roberts's recognition of that interest appears in the foreword to the book's third edition, where Roberts mentioned the then prince-president's order to copy and circulate the book.

33. Frégier, *Des classes dangereuses de la population dans les grandes villes*, 144–45.

34. Lévy, *Traité d'hygiène publique et privée*, 555.

35. Alphonse du Breuil, *Cours élémentaire théoretique et pratique d'arboriculture, comprenant l'étude des pépinières d'arbres et d'arbrisseaux forestiers, fruitiers et d'ornement; celle des plantations d'alignement forestières et d'ornement; la culture spéciale des arbres à fruits, à cidre et de ceux à fruits de table* (Paris: Langlois et Leclercq, 1853), 2.

36. Tardieu, *Dictionnaire d'hygiène publique et de salubrité*, 113.

37. Édouard André, *L'art des jardins: Traité général de la composition des parcs et jardins* (1879; rpt. Marseille: Lafitte reprints, 1983), 103.

38. Jean-Baptiste Boussingault, *Agronomie, chimie agricole et physiologie, tome deuxième* (Paris: Mallet-Bachelier, 1861), 317.

39. Ibid., 318.

40. Sabine Barles and Laurence Lestel, "The Nitrogen Question: Urbanization, Industrialization, and River Quality in Paris, 1830–1939," *Journal of Urban History* 33, no. 5 (July 2007): 796. It should be noted that all of the early nineteenth-century studies of plant growth and what is now known as "the nitrogen cycle" occurred in the absence of a full understanding of the chemistry of photosynthesis, or the way in which bacteria in the soil operated on nitrogen compounds to facilitate plant absorption.

41. See Richard P. Aulie, "Boussingault and the Nitrogen Cycle," *Proceedings of the American Philosophical Society* 114, no. 6 (December 1970): 446–47.

42. Michel-Eugène Chevreul, *Recherches expérimentales sur la végétation par M. Georges Ville: Absorption de l'azote de l'air par les plantes* (Paris: Imprimerie de la Martinet, 1855), 6.

43. Alexandre Jouanet, *Paris et ses plantations* (Paris: Typographie de Morris et cie., 1860), 9.

44. Fonssagrives, *Hygiène et assainissement des villes*, 173–74. It is worth noting that Adolphe Alphand and key members of his team—Jean Darcel, Jean-Pierre Barillet-Deschamps, and Édouard André—all lived and worked in Bordeaux in 1847 precisely at the time when Jeannel was presenting his ideas to the *Société de médecine de Bordeaux* challenging the orthodox opinion about tree transpiration and biology. Georges-Eugène Haussmann arrived a year later from nearby Blaye.

45. Ibid., 174 (emphasis in original).

46. Ibid., 175.

47. Jean-François Jeannel, "Des plantations d'arbres dans l'intérieur des villes au point de vue de l'hygiène publique, conférence faite au Jardin d'acclimatation, le 23 juillet 1872," *Bulletin mensuel de la Société d'acclimatation,* August 1872, 532.

48. Ibid., 533.

49. Ibid., 535.

50. Ibid., 538.

51. Georges-Eugène Haussmann, *Mémoires du Baron Haussmann* 1: *Grands Travaux de Paris* (Paris: Guy Durier, 1979), 240.

52. Ibid.

53. Ibid., 226.

54. "Travaux de Paris: Mémoire présenté par M. le Sénateur Préfet de la Seine au Conseil général du département et au Conseil municipal de la Ville de Paris," *Revue générale d'architecture et travaux publics* 24 (1866): 263 (emphasis in original).

55. Ibid., 226.

56. AP, TriBriand VM 19, Ville de Paris, Service des Promenades et Plantations, *Tableau statisique des bois, parcs, squares, promenades publiques plantées,* 28 August 1868.

57. Carmona, *Haussmann,* 451–52.

58. Adolphe Alphand and Alfred Auguste Ernouf, *L'art des jardins. Parcs—jardins—promenades—étude historique—principles de la composition des jardins—plantations—décoration pittoresque et artistique des parcs et jardins publics; traité pratique et didactique* (Paris: J. Rothschild, 1886), 325–26.

59. For pronatalism and public policy, see Joshua Cole, *The Power of Large Numbers: Population, Politics, and Gender in Nineteenth-Century France* (Ithaca, N.Y.: Cornell University Press, 2000); Elinor Accampo, Rachel Fuchs, and Mary Lynn Stewart, *Gender and the Politics of Social Reform in France, 1870–1914* (Baltimore: Johns Hopkins University Press, 1995); and Fuchs, *Poor and Pregnant in Paris,* chap. 3.

60. Adolphe Trebuchet, "Report on the labours of the 'Conseil de Salubrité' of Paris, from 1829–1839, by M. Trebuchet," in Edwin Chadwick, *Report to Her Majesty's Principal Secretary of State,* 415.

61. Lévy, *Traité d'hygiène publique et privée,* 593.

62. Ibid.

63. Alfred Donné, *Conseils aux mères sur la manière d'élever les enfans nouveau-nés, ou de l'éducation physique des enfants du premier âge, 1842,* qtd. in Lévy, *Traité d'hygiène publique et privé,* 594 (emphasis added).

64. Louis Cyprien Descieux, *Leçons d'hygiène à l'usage des enfants des écoles primaires* (Paris: Imprimerie et librairie classiques, 1858). For later editions, see National Library of Medicine (U.S.), *Index-Catalogue of the Library of the Surgeon-General's Office* (Washington, D.C.: U.S. Dept. of Health, Education, and Welfare, Public Health Service [etc.], U.S. Govt. Printing Office, 1880); and National Library of Medicine (U.S.) and Carnegie Institution (Washington, D.C.), *Index Medicus: A Monthly Classified Record of the Current Medical Literature of the World* (Washington, D.C.: U.S. Govt. Printing Office, 1880), vol. 2.

65. Descieux, *Leçons d'hygiène,* 7.

66. Ibid., 51–52.

67. Ibid., 53–54.

68. Louis Cyprien Descieux, *Manuel d'hygiène à l'usage des élèves des écoles normales primaires, des écoles spéciales, des lycées, collèges et séminaires* (Paris: P. Dupont, 1867), 107.

69. Ibid., 110.

70. Ibid., 111.

71. See Cole, *The Power of Large Numbers,* chap. 5; and Fuchs, *Poor and Pregnant in Paris,* 56–67.

72. Jean-Baptiste Fonssagrives, *L'éducation physique des jeunes filles; ou, avis aux mères sur l'art de diriger leur santé et leur développement* (Paris: L. Hachette, 1869), 72.

73. Cole, *The Power of Large Numbers,* 178–79.

74. Fonssagrives, *L'éducation physique des jeunes filles,* 73.

75. Ibid., 74.

76. *Paris guide par les principaux écrivains et artistes de la France. Deuxième partie: La vie* (Paris: A. Lacroix, Verboeckhoven et cie., 1867), 602.

77. Ibid., 603.

78. William Robinson, *The Parks, Promenades, and Gardens of Paris Described and Considered in Relation to the Wants of Our Own Cities and of Public and Private Gardens* (London: J. Murray, 1869), 87–88.

CHAPTER THREE

Epigraph: AP, VONC 908, *Registre du Garde Aubert,* garde portier du Bois de Boulogne, 1 August 1853.

1. AP, VONC 1067, *Ordre de Service pour les gardes du Bois de Boulogne,* signé A. Alphand, Ponts et Chaussées, Département de la Seine, Ville de Paris, Service municipal des Promenades et Plantations, 12 September 1856.

2. Ibid.

3. Ibid.

4. Paul Joanne, ed., *Dictionnaire géographique et administratif de la France* (Paris: Librarie Hachette et cie., 1898), 3189.

5. AP, Perotin 10653 107, *Ordre de Service pour les gardes du Bois de Boulogne, Les attributions des gardes,* Ponts et Chaussées, Département de la Seine, Ville de Paris, Service municipal des Promenades et Plantations, September 1856. This document, signed by Alphand, shows where, using last names, each individual guard was to be stationed in the park. There are names crossed out and moved, so it appears to have been written at a time when assignments were still being determined.

6. See for example AP, Perotin 10653 80, *Rapport du garde, Square Monge,* Département de la Seine, Direction des Travaux de Paris, Promenades intérieures, 16 October 1874; and AP, VM90 5, *Rapport du Garde Gras,* Service de la Place des Vosges, République Française, Département de la Seine, Ville de Paris, Direction administrative des Service d'Architecture et des Promenades et Plantations, Service des Promenades, Parc de Montsouris, 6 July 1901.

7. AP, VONC 908, *Registre du Garde Aubert,* garde portier du Bois de Boulogne, 1853–1856; AP, VONC 881, *Registre de Garde Labadie,* garde forestier à la résidence Longchamps, 1861–1863; and AP, VONC 1067 *Registre de Garde Bize,* chargé de la surveillance au [illegible] du lac supérieur, 1856–1860.

8. AP, VONC 881, *Registre de Garde Labadie.* See for example daily entries for 21 October 1861 and 6 October 1862.

9. Ibid.; and Joanne, *Dictionnaire géographique et administratif de la France,* 3189.

10. AP, VONC 1067, *Ordre de Service,* 12 September 1856.

11. AP, VONC 908, *Registre du Garde Aubert;* AP, VONC 881, *Registre de Garde Labadie;* and AP, VONC 1067 *Registre de Garde Bize.*

12. Adolphe Alphand, *Les promenades de Paris: Histoire—description des embellissements—dépenses de création et d'entretien des Bois de Boulogne et de Vincennes, Champs-Élysées—Parcs—squares—boulevards—places plantées, Étude sur l'art des jardins et arboretum, 487 gravures sur bois, 80 sur acier, 23 chromolithographies* (1867; rpt. Princeton, N.J.: Princeton Architectural Press, 1984), 72–73.

13. AP, Perotin 10653 107, *Rapport de l'Ingénieur ordinaire sur la demande de Sieur* [illegible] *et compagnie tendant à être autorisés à fournir du Gaz portif pour l'éclairage de divers établissement du Bois de Boulogne,* Pont et Chaussées, Département de la Seine, Ville de Paris, Service spécial des Promenades et Plantations, 11 June 1856.

14. AP, VONC 881, *Registre de Garde Labadie.*

15. Ibid.

16. AP, VONC 1067 *Ordre du jour* signé Alphand, 21 June 1858, in *Registre de Garde Bize.*

17. AP, VONC 1067, *Ordre de Service pour les gardes du Bois de Boulogne,* 12 September 1856.

18. Ibid.

19. AP, VONC 195, *Instructions sur la marche des services des promenades et de l'éclairage, rédigée par M. l'Ingénieur en Chef Darcel, au moment ou ces services ont été pris par M.M. les ingénieurs de section,* [1867].

20. AP, VONC 195, *Note de Service,* signée Alphand, 21 July 1874.

21. Ibid.

22. AP, Perotin 10653 90, *Copie d'une circulaire plus spéciale aux bois* [sic] *de Boulogne et Vincennes,* Ponts et Chaussées, Département de la Seine, Ville de Paris, Direction des Travaux de Paris, Promenades, Éclairage, Voitures et Concessions, 24 October 1874.

23. Ibid.

24. Ibid.

25. William A. Peniston, *Pederasts and Others: Urban Culture and Sexual Identity in Nineteenth-Century Paris* (New York: Harrington Park Press, 2004), 35–42; Julian Jackson, *Living in Arcadia: Homosexuality, Politics, and Morality in France from the Liberation to AIDS* (Chicago: University of Chicago Press, 2009), 21–31.

26. Peniston, *Pederasts and Others*, 42; Archives de la Préfecture de Police de Paris (hereafter APPP), BB6, *Registre "Pédérastes et divers,"* shows police sweeps and arrests involving various passages, boulevard urinals, and train station restrooms. As for the parks and squares, the bushes and urinal near the Café des Ambassadeurs in the Champs-Élysées gardens, the area of the Bois de Vincennes near the fort, and the Allée des Creuzes "dites Soupirs" (so-called Allée of Sighs) in the Parc Monceau appear to have been popular meeting places known to the police and sites of surveillance and arrests.

27. *Édifices de Paris construits de 1871 à 1889,* Institute Géographique Nationale, 8286, PL XVI. A copy of this map from the collection of the Bibliothèque Historique de la Ville de Paris was presented to me as a gift.

28. AP, 2380W 28, *Square des Invalides, Plainte de Général Brincourt, Garde Moreaux,* Rapport de l'inspecteur, Ponts et Chaussées, Département de la Seine, Ville de Paris, Direction de la Voie publique et des Promenades, Service des Promenades et Plantations, 20 June 1870.

29. Ibid.

30. Ibid.

31. Ibid.

32. Ibid.

33. Ibid.

34. Francois Loyer and Stan Neumann, *Paris, Roman d'une ville* (Ho-Ho-Kus, N.J.: Roland Collection, 1990).

35. AP, Perotin 10653 90, *Squares: Observations divers sur plusieurs abus ou négligence, Sieur Démophile,* Rapport de Conducteur principal Inspecteur, Ponts et Chaussées, Département de la Seine, Ville de Paris, Direction de la Voie publique et des Promenades, Service des Promenades et Plantations, 20 August 1873.

36. Ibid.

37. Ibid.

38. Ibid.

39. AP, 2380 40, Rapport du Garde Millet, Square de la Chapelle, République Française, Direction de Travaux de Paris, Promenades intérieures, 21 August 1889.

40. Ibid., 5 June 1889.

41. Ibid.

42. Ibid.

43. AP, 1304W 210, Rapport du Garde Grad, Square Montholon, République Française, Direction de Travaux de Paris, Promenades intérieures, 9 August 1888.

44. Ibid.

45. Ibid.

46. AP, VM90 5, Letter from the Director of the École communale des Garçons to M. le Maire de 4e arrondissement, 4 December 1892 (emphasis in original).

47. Ibid.
48. Ibid.
49. Ibid.
50. Ibid.
51. AP, VM90 5, *Plaint contre le gardien du square de la Place des Vosges,* République Française, Préfecture du Département de la Seine, Direction de l'Enseignement primaire, 28 January 1893.
52. Ibid.
53. See for example AP, Perotin 10653 90, *Squares: Observations divers sur plusieurs abus ou négligence,* 20 August 1873.
54. AP, 1304W 39, *Dossiers individuels de personnel du Service d'Architecture, Cantonniers,* Ponts et Chaussées, Département de la Seine, Direction des Travaux de Paris, Service des Promenades et Plantations.
55. AP, Perotin 10653 62, *Service dans Paris,* signé G. E. Haussmann, Préfecture du Département de la Seine, Promenades et Plantations, 29 February 1856.
56. AP, Perotin 10653 247, *Uniforme des Cantonniers,* Circulaire aux conducteurs de M. Grégoire, Ingénieur ordinaire, 31 January 1860; AP, Perotin 10653 247, *État indiquant les effets d'uniforme donnés gratuitement par l'administration aux cantonniers des cimetières, le 19 Aout 1860,* Ponts et Chaussées, Service municipal de Paris, Service des Promenades et Plantations.
57. AP, Perotin 10653 247, *Habillement instructions, Demande de renseignements,* Ponts et Chaussées, Département de la Seine, Ville de Paris, Service municipal des Travaux Publics, April 1863.
58. Carol Harrison, *The Bourgeois Citizen in the Nineteenth-Century France: Gender, Sociability, and the Uses of Emulation* (Oxford, U.K.: Oxford University Press, 1999), 129–30.
59. AP, 1304W 39, *Dossiers individuels de personnel du Service d'Architecture, Cantonniers, Chareille, Jean,* Ponts et Chaussées, Département de la Seine, Direction des Travaux de Paris, Service des Promenades et Plantations.
60. AP, 1304W 39, *Dossiers individuels de personnel du Service d'Architecture, Cantonniers, Farrett, Étienne,* Ponts et Chaussées, Département de la Seine, Direction des Travaux de Paris, Service des Promenades et Plantations; and *Dossiers individuels de personnel du Service d'Architecture, Cantonniers, Charielle, Jean.*
61. On the culture of the *Corps des Ponts et Chaussées,* see Lenard R. Berlanstein, *Big Business and Industrial Conflict in Nineteenth-Century France: A Social History of the Parisian Gas Company* (Berkeley: University of California Press, 1991), 97–107; for Alphand's military service and general attitudes among military commanders toward the Communards, see Massillon Rouvet, *Viollet-le-duc et Alphand au siège de Paris* (Paris: Librairies-Imprimeries Réunies, 1892), 195–204, and Robert Tombs, *The Paris Commune, 1871* (London: Longman, 1999), 158–61, 179.
62. AP, 1304W 39, *Dossiers individuels de personnel du Service d'Architecture, Cantonniers, Flammant, Louis,* Ponts et Chaussées, Département de la Seine, Direction des Travaux de Paris, Service des Promenades et Plantations.
63. AP, 1304W 39, *Dossiers individuels de personnel du Service d'Architecture, Cantonniers, Jouas, Jean,* Ponts et Chaussées, Département de la Seine, Direction des Travaux de Paris, Service des Promenades et Plantations.
64. See AP, VM90 5, *Square Montholon, Demande d'autorisation de faire circuler des voitures attelées de chèvres,* Rapport du conducteur adjoint et du jardinier principal, République Française,

Département de la Seine, Ville de Paris, Direction administrative des Service d'Architecture et des Promenades et Plantations, 28 August 1902; AP, 1304W 159, *Square Avenue de la République, Demande d'autorisation de vendre des fleurs, M. Belotte, Charles,* Rapport de l'inspecteur, République Française, Ponts et Chaussées, Direction des Travaux de Paris, Service de la Voie publique et des Promenades. 27 November 1890; AP, VM90 5, *Le Sr. Anicetto demande d'être autorisé à établir un théâtre de marionnettes, Place des Vosges,* Rapport du conducteur principal, Préfecture du Département de la Seine, Direction des Travaux de Paris, Service des Promenades et Plantations, 5 April 1873; AP, VONC 225, *Installation des appareils stéréoscopes,* Rapport de l'inspecteur, République Française, Ponts et Chaussées, Direction des Travaux de Paris, Service de la Voie publique et des Promenades, 1894. Two photographs accompany this last report, showing the cast iron stereoscopes. Park visitors paid ten centimes to view a photographic image inside the box. In the case of the photographs attached to the report, the images were of the Pyrenees and the obsequies of President Carnot.

65. AP, 1034W 113, *Parc des Buttes Chaumont, Café Brasserie, Demande de délai de paiement de loyer, le Sieur Bridoux,* Rapport du conservateur, Ponts et Chaussées, Département de la Seine, Ville de Paris, Directions de la Voie publique et des Promenades, Service des Promenades et Plantations, 28 November 1872; AP, 1304W 113, Parc des Buttes Chaumont, *Café Brasserie Puebla, Demande de l'autorisation de travaux, M. Bouquet Concessionaire,* Rapport de l'architecture, République Française, Ponts et Chaussées, Direction des Travaux de Paris, 22 October 1885.

66. AP, VM90 2, *Permission, Bois de Vincennes, Vente de plaisir, Mme Vve Vinel, 6 rue Charonne, à Saint-Mandé,* Rapport du conservateur, République Française, Préfecture du Département de la Seine, Ville de Paris, Direction administrative des Services d'Architecture et des Promenades et Plantations, Service des Promenades et Plantations, 30 July 1901.

67. See AP, Perotin 10253 54, Dossier D, *Bois de Boulogne, Patinage,* Préfecture du Département de la Seine, Letter from M. le Préfet de la Seine to M. le Chef du Service du Promenades et Plantations, 14 January 1857; "Types Parisiens, No. 6, Le charmeur d'oiseaux," *Le Monde illustré* 11, no. 287 (11 October 1862): 231; and "Le charmeur d'oiseaux," *La Rue: Paris pittoresque et populaire,* 8 June 1867. Bird charmers were a kind of street performer most common in the Tuileries but appeared later in the other gardens. These men and women amazed visitors by seeming to call flocks of birds to themselves, allowing them to rest on their head and shoulders.

68. AP, VONC 908, *Conditions à imposer aux louers de chevaux et ânes, ordre du jour* in *Registre du Garde Aubert,* 2 July 1854.

69. Ibid.

70. AP, 1304W 113, *Parc des Buttes Chaumont, Pavillon Puebla, Demande d'autorisation de laisser jouer piano, M. Bouquet,* Rapport du Conducteur, Ponts et Chaussées, Département de la Seine, Ville de Paris, Direction de la Voie publique et des Promenades, Service des Promenades et Plantations, 18 September 1876.

71. For a discussion of the political nature of entertainments (particularly song) in the café concerts during the Second Empire, see T. J. Clark, *The Painting of Modern Life* (Princeton, N.J.: Princeton University Press, 1984.)

72. AP, 1304W 113, *Parc des Buttes Chaumont, Pavillon Puebla, Demande d'autorisation de laisser jouer piano, M. Bouquet,* 18 September 1876.

73. AP, 1304W 113, *Café Brasserie Puebla, Demande de l'autorisation de travaux, M. Bouquet, Concessionaire,* 22 October 1885.

74. Ibid.

75. AP, 1304W 113, *Note de Service,* Département de la Seine, Direction des Travaux de Paris, Promenades intérieures, 28 January 1876.

76. AP, 1304W 184, *Parc des Buttes Chaumont, Location au Srs. Manceau et Bridoux, du pavillon à usage de Café Restaurant et Café Brasserie,* Préfecture du Département de la Seine, Direction des Affaires municipales, 28 February 1868.

77. Ibid.

78. AP, VONC 291, *Parc des Buttes Chaumont, Demande de la femme Laurens, tendant à d'obtenir l'autorisation d'établir une cantine sur le chantier,* Rapport de l'ingénieur ordinaire, Ponts et Chaussées, Département de la Seine, Ville de Paris, Direction de la Voie publique et des Promenades, Service des Promenades et Plantations, 29 May 1865.

79. AP, 2380W 20, *Parc de Montsouris, Demande de séjour pour photographier, M. Monsus, pétitionnaire,* Rapport du conducteur, République Française, Département de la Seine, Ville de Paris, Direction administrative des Service d'Architecture et des Promenades Plantations, Service technique des Promenades et Plantations, 12 May 1908.

80. AP, 1304W 114, *Parc des Buttes-Chaumont, Demande de Concession, M. Bouquet,* Rapport l'architecte des promenades, République Française, Ponts et Chaussées, Direction des Travaux de Paris, Service des Promenades, 17 June 1888.

81. Ibid.

82. See for example the following documents concerning the situation in the Square de l'Archevêché described in AP, Perotin 10653 110, *Square l'Archevêché, Suppression du garde,* Ponts et Chaussées, Département de la Seine, Ville de Paris, Direction de la Voie publique et des Promenades, Service des Promenades et Plantations, 21 September 1870; *Square l'Archevêché, Abatage d'arbres par l'artilleurs de la garde nationale,* Ponts et Chaussées, Département de la Seine, Ville de Paris, Direction de la Voie publique et des Promenades, Service des Promenades et Plantations, 20 January 1871; and *Square l'Archevêché, Affectation au Parc d'Artillerie de la garde nationale, Approbation de Dépense,* République Française, Mairie de Paris, Directions de la Voie publique et des Promenades, 24 February 1871.

83. AP, 1304W 113, *Parc des Buttes Chaumont, Sieur Bouquet, Demande de dégrèvement de loyer,* Rapport du Conservateur, Ponts et Chaussées, Département de la Seine, Ville de Paris, Directions de la Voie publique et des Promenades, Service des Promenades et Plantations, 26 August 1871.

84. AP, 1304W 113, contains a series of yearly license lists (1876–92) compiled by Ponts et Chaussées, Direction des Travaux de Paris, titled *État des sommes verser dans la caisse de la Ville de Paris pour permissions de pêche dans le lac du parc des Buttes Chaumont* (sometimes titled *État des sommes recouvrer par la Ville de Paris pour permissions de pêche dans le lac du parc des Buttes Chaumont*).

85. AP, 1034W 113, *Parc des Buttes Chaumont, Café Brasserie, Demande de delai de paiement de loyer, le Sieur Bridoux,* Rapport du conservateur, Ponts et Chaussées, Département de la Seine, Ville de Paris, Directions de la Voie publique et des Promenades, Service des Promenades et Plantations, 28 November 1872.

86. David Harvey, *Paris: Capital of Modernity* (New York: Routledge, 2003), 329.

87. AP, 1034W 113, *Parc des Buttes Chaumont, Café Brasserie, Demande de delai de paiement de loyer, le Sieur Bridoux,* 28 November 1872.

88. AP, 1304W 159, *Square Avenue de la République, Demande d'autorisation de vendre des fleurs, M. Belotte, Charles,* Rapport de l'inspecteur, République Française, Ponts et Chaussées, Direction des Travaux de Paris, Service de la Voie publique et des Promenades, 27 November 1890.

89. AP, VM90 5, *Square de la Place des Vosges, Demande de concession Mme Vve Épistalié,* Rapport du conducteur, République Française, Département de la Seine, Ville de Paris, Direction administrative des Services d'Architecture et des Promenades et Plantations, Service technique des Promenades et Plantations, 23 April 1902.

90. The archival record is replete with *arrêtés* and decisions concerning women and these kinds of small concessions in greenspaces in every part of the city. For some examples, see: AP, 2380W 28, *Stationnement dans les kiosques du jardin des Invalides,* Préfecture du Département de la Seine, Administration de la Ville de Paris, Direction des Affaires municipales, Arrêté signé G. E. Haussmann, 23 January 1869; and AP, VM90 5, *Square des Vosges, Kiosque de Mme Felten, Demande de déplacement, M. Faucheux,* République Française, Ponts et Chaussées, Direction des Travaux de Paris, 22 May 1885.

91. Victoria Thompson, *The Virtuous Marketplace: Women and Men, Money and Politics in Paris, 1830–1870* (Baltimore: Johns Hopkins University Press, 2000), chap. 3.

92. AP, 1304W 210, Letter from Monsieur T. P—— to Monsieur l'inspecteur, 28 February 1878. The surname is illegible.

93. Ibid.

94. AP, 1304W 210, Rapport du garde brigadier Grivillers, Département de la Seine, Direction des Travaux de Paris, Promenades intérieures, 4 March 1878 (emphasis added here to highlight the brigadier's use of the past tense contained in the original document, which indicates an insistence that this occurred only once).

95. Ibid.

96. Ibid.

97. AP, 1304W 174, *Square Louis XVI, Plainte contre la receveuse des chaises, Mmes Trouble et Ullmann,* Rapport de l'inspecteur, République Française, Ponts et Chaussées, Direction des Travaux de Paris, 28 July 1885.

98. Ibid.

99. AP, 1304W 174, *Square Louis XVI, Plainte contre la personne chargée de la perception des siéges,* Rapport de l'inspecteur, République Française, Ponts et Chaussées, Direction des Travaux de Paris, Service de l'Ingénieur en Chef d'Éclairage des Promenades et des Concessions sur la Voie publique, 10 June 1886.

100. Ibid.

101. AP, VM90 5, Letter from Adolphe Alphand, *pour le Préfet et par autorisation,* to M. Faucheux, rue des Vosges 12, 21 July 1884.

102. AP, VM90 5, Letter from Nicolas Felten to Monsieur le Directeur [des Travaux de Paris], 13 August 1884.

103. AP, VM90 5, Letter from "les soussignés" to the Administration municipale (no addressee), 23 July 1884.

104. AP, VM90 5, *Square des Vosges, Kiosque de Mme Felten, Demande de déplacement, M. Faucheux,* République Française, Ponts et Chaussées, Direction des Travaux de Paris, Service des Promenades, 22 May 1885.

105. Ibid. (emphasis in original).
106. AP, VM90 5, Letter from E. Roche to Monsieur le Directeur des Ponts et Chaussées, 7 September 1886.
107. Ibid.
108. Ibid.

CHAPTER FOUR

Epigraph: AP, VM90 5, Letter of complaint signed A. Pichot, *Instituteur libre* to a local municipal councilor, 4 August 1877.

1. Victor Turner, *The Ritual Process: Structure and Anti-structure* (Chicago: Aldine Publishing Co., 1969), 131–95.
2. Ibid., 96.
3. Ibid., 131–95.
4. John Gage, *Color and Culture: Practice and Meaning from Antiquity to Abstraction* (Berkeley: University of California Press, 1999), 173–76.
5. Michel Carmona, *Haussmann* (Paris: Fayard, 2000), 568–69; David Pinkney, *Napoleon III and the Rebuilding of Paris* (Princeton, N.J.: Princeton University Press, 1958), 121.
6. AP, 1304W 59, *Square Louis XVI; Réclamation des propriétaires voisins,* Rapport de l'architecte en chef, Département de la Seine, Ville de Paris, Ponts et Chaussées Service municipal des Travaux Publics, Service des Promenades et Plantations, 3 July 1865 (emphasis in original).
7. Archive de Paris VONC 291, *Parc des Buttes Chaumont, Demande des habitants de la rue des Allouettes* [sic] *tendant à obtenir l'ouverture d'une entrée au parc vis à vis cette* [ill.] *voie,* Rapport de l'Ingénieur ordinaire Darcel, Département de la Seine, Ville de Paris, Ponts et Chaussées, Service municipal des Travaux Publics, Service des Promenades et Plantations, 30 August 1866.
8. Ibid.
9. Adolphe Alphand, *Les promenades de Paris: Histoire—description des embellissements—dépenses de création et d'entretien des Bois de Boulogne et de Vincennes, Champs-Élysées—Parcs—squares—boulevards—places plantées, Étude sur l'art des jardins et arboretum, 487 gravures sur bois, 80 sur acier, 23 chromolithographies* (1867; rpt. Princeton, N.J.: Princeton Architectural Press, 1984), n. p.
10. AP, 1304W 118, *Parc des Buttes Chaumont: Demande d'ouverture d'une nouvelle porte—Pétition,* Rapport du conducteur principal, chef de Service, Préfecture de la Seine, Ponts et Chaussées, Direction de la Voie publique et des Promenades, Service des Promenades et Plantations. 21 December 1879.
11. James D. McCabe Jr., *Paris by Sunlight and Gaslight: A Work Descriptive of the Mysteries and Miseries, the Virtues, the Vices, the Splendors and the Crimes of the City of Paris* (Philadelphia: National Publishing Co., 1869), 262.
12. On a present-day map of Paris, that former route would have connected rue de la Villette in the south with rue d'Hautpoul in the north.
13. AP, VONC 291, *Parc des Buttes Chaumont, Observations de la rue de la Villette relatives à la suppression d'une section de cette voie,* Rapport de l'ingénieur ordinaire, signé Darcel, Département

de la Seine, Ville de Paris, Ponts et Chaussées Service municipal des Travaux Publics, Service des Promenades et Plantations, 12 February 1866.

14. The Parisian press had been reporting extensively on the public's fight to prevent alteration to the Luxembourg Garden and the role the emperor played in that struggle. See Richard S. Hopkins, "*Sauvons le Luxembourg*: Urban Greenspace as Private Domain and Public Battleground, 1865–1867," *Journal of Urban History* 37, no. 1 (January 2011): 43–58. As for Buttes Chaumont, the emperor made several tours of the work site, questioning engineers and workers about the progress, something observant residents would have been aware of.

15. AP, VONC 291, *Parc des Buttes Chaumont, Plainte adressée à l'Empereur par les habitants des rues du Plateau et des Allouettes* [sic], Rapport de l'ingénieur ordinaire, signé Darcel, Département de la Seine, Ville de Paris, Ponts et Chaussées Service municipal des Travaux Publics, Service des Promenades et Plantations, 5 February 1867.

16. AP, 1304W 155, Dossier 7, *Square Parmentier, Ouverture d'une nouvelle porte sur la rue Guilhem, Pétition des divers habitants,* Rapport du Conducteur principal, Département de la Seine, Ville de Paris, Ponts et Chaussées, Service municipal des Travaux Publics, Service des Promenades et Plantations, 13 May 1876.

17. AP, 1304W 118, *Parc des Buttes Chaumont,* Rapport du sous-ingénieur, République Française, Ponts et Chaussées, Direction des Travaux de Paris, Service des Promenades, 18 April 1883.

18. Ibid.

19. AP, 2380W 20, *Projet d'installation du clôture et d'éclairage pour le sentier reliant la station de Sceaux-ceinture à la rue Gazan dans le Parc de Montsouris, Plan des Travaux,* République Française, Ponts et Chaussées, Direction des Travaux de Paris, Service des Promenades, 23 January 1882; AP, 2380W 20, *Parc de Montsouris, Installation des barrières et d'appareils d'éclairage dans le sentier reliant la rue Gazan à la station de Sceaux-Ceinture, croquis,* n.d.

20. AP, 2380W 20, *Park Montsouris, Passage reliant les gares de la Glacière et Sceaux-ceinture. Éclairage de nuit,* Rapport des l'ingénieurs ordinaire, signé Seilheimer, République Française, Ponts et Chaussées, Direction des Travaux de Paris, Service des Promenades, 12 October 1882.

21. AP, 1304W 118, *Park des Buttes Chaumont, Travaux à exécuter par le Service municipal au compte des particuliers,* République Française, Ponts et Chaussées, Direction des Travaux de Paris, Service des Promenades, June 1890; AP, 1304W 118, *Projet d'ouverture d'un guichet dans la grille de clôture du Parc des Buttes Chaumont, Rue Manin entre la porte Laumière et celle de Crimée, Plan,* République Française, Ponts et Chaussées, Direction des Travaux de Paris, Service des Promenades, July 1890.

22. AP, 1304W 118, *Park des Buttes Chaumont, Demande d'ouverture d'une porte avec sentier, Grébauval, Conseilleur Municipal,* Rapport de l'inspecteur, République Française, Préfecture de la Seine, Direction administrative des Travaux de Paris, Service d'éclairage, des Promenades et des Concessions, Inspection des Promenades, 22 June 1893.

23. AP, 1304W 118, *Park des Buttes Chaumont, Ouverture d'une porte entre les rues Botzaris et Crimée, Proposition de MM Grébauval et Charles Bos, Conseillers Municipaux,* Rapport du jardinier en chef, République Française, Préfecture de la Seine, Direction administrative des Travaux de Paris, Service des Promenades, 29 August 1896.

24. Ibid.

25. AP, VM90 5, Lettre de M. l'Inspecteur Général à M. le Maire du 4e arrondissement, 26 May 1887.

26. AP, VONC 69, *Rapports sur les fêtes foraines es sur les dangers qu'elles presenter pour la santé public,* Département de la Seine, Conseil d'hygiène, 16 February 1887.

27. AP, 1304W 184, Dossier 3, Petition to the Conseil municipal made by members of the *Société laïque d'appui fraternel,* 9 July 1884.

28. AP, VM90 5, *Square Place des Vosges, Demande d'autorisation de donner des concerts.* M. Royer, Rapport de l'inspecteur, République Française, Ponts et Chaussées, Direction des Travaux de Paris, Service des Promenades, 13 April 1887.

29. AP, VM90 5, *Square Place des Vosges, Demande d'autorisation des concerts, La Société «La Liberté,»* Rapport de l'inspecteur ordinaire, République Française, Ponts et Chaussées, Direction des Travaux de Paris, Service des Promenades, 21 May 1887.

30. AP, 1304W 155, Rapport du Garde Géraud, République Française, Département de la Seine, Ville de Paris, Direction administrative des Service d'Architecture et des Promenades et Plantations, Service des Promenades, Service du Square Parmentier, 21 June 1901.

31. AP, 2380 20, Rapport du Garde, République Française, Département de la Seine, Ville de Paris, Direction Administrative des Service d'Architecture et des Promenades et Plantations, Service des Promenades, Parc de Montsouris, 24 May 1903.

32. Ibid. For more on the popularity of military concerts in parks and squares, see also AP, 1304W 162, *Square du Temple; Demande de création d'un emplacement pour donner des concerts militaires, Pétition collective,* Rapport de l'inspecteur. République Française, Ponts et Chaussées, Direction des Travaux de Paris, Services de la Voie publique et des Promenades, 7 July 1892; AP, TriBriand 118, *Parc de Montsouris, Pétition demandant l'organisation des concerts militaires le dimanche et le jeudi,* Rapport de l'inspecteur, République Française, Préfecture de la Seine, Direction administrative des Travaux de Paris, Voirie et Travaux d'Ingénieurs, Service technique des Promenades, 5 December 1895; and AP, TriBriand 118, Letter from le Général Saussier, Gouverneur Militaire de Paris to M. le Préfet de la Seine, 31 December 1895.

33. AP, 1304W 167, Dossier 16, *Jardin de Luxembourg, Demande d'arrosage, Plainte de M. Marceaux,* Rapport du conducteur, Département de la Seine, Ville de Paris, Ponts et Chaussées, Direction des Travaux de Paris, Service des Promenades, 17 December 1872.

34. Ibid.

35. Ibid.

36. See for example the following instances in which requests by persons of high military and civilian status were nevertheless declined by the park service based on the content of the petition. A request made by a former naval lieutenant and rejected is contained in AP, 1304W 114, *Parc des Buttes Chaumont, Autorisation demandé par M. LaBrousse pour y établir des régates et des course vélocipèdes,* Rapport de l'ingénieur en chef, Département de la Seine, Ville de Paris, Ponts et Chaussées, Direction de la Voie publique et des Promenades, Service des Promenades, 21 April 1868; and a request made by the Peruvian ambassador to France and rejected, is contained in AP, 10653 74, *Bois de Boulogne, Établissement d'un jeu de cricket, Note à M. le Chef du Cabinet,* Préfecture du Département de la Seine, Direction des travaux de Paris, 9 April 1878.

37. AP, 1304W 118, *Parc des Buttes Chaumont, Plainte de M. Buchet au sujet du mode de perception des sièges et de l'arrosage de la promenade, envoi par M. Grébauval Conseiller municipal,* Rapport de l'inspecteur, République Française, Préfecture de la Seine, Direction administrative des Travaux de Paris, Service technique des Promenades, Inspection des Promenades, 23 June 1896.

38. As part of the contractual relationship between the concessionaires and the park service, the administration had the right to approve all hires and could call for the dismissal of any person employed within the park boundaries.

39. AP, 1304W 118, *Parc des Buttes Chaumont, Plainte de M. Buchet au sujet du mode de perception des sièges et de l'arrosage de la promenade,* 23 June 1896.

40. AP, 1304W 118, *Park de Montsouris, Réclamation de M. Baronnier,* Rapport du conducteur, République Française, Préfecture de la Seine, Direction administrative des Service d'Architecture et des Promenades et Plantations, Service technique des Promenades et Plantations, 16 December 1903.

41. Ibid.

42. AP, 1304W 210, *Square Montholon, Suppression de la pièce d'eau,* Rapport du Conducteur principal et inspecteur, Département de la Seine, Ville de Paris, Ponts et Chaussées, Direction de la Voie publique et des Promenades, Service des Promenades, 30 October 1873.

43. Ibid.

44. AP, VM90 5, *Square Montholon, Suppression de la pièce d'eau,* Mémoire au Conseil Municipal signé M. le Préfet de la Seine, 4 December 1873.

45. AP, VM90 5, Petition addressed to M. Alphand from the riverains of the Square Montholon, 7 March 1874.

46. Ibid.

47. AP, VM90 5, Petition addressed to M. le Préfet de la Département de la Seine from riverains of the Square Montholon, signed "vos respecteux serviteurs," 16 March 1874 (emphasis in original).

48. AP, 1304W 210, Dossier 8, *Projet de modification à apporter à une partie du Square Montholon,* République Française, Ponts et Chaussées, Direction des Travaux de Paris, Service des Promenades, 1 December 1886. The removal of the water feature as planned in 1874 is confirmed by this document containing the design of a subsequent alteration in 1886.

49. AP, TriBriand 118, *Parc de Montsouris; Suppression de pelouses, Arrêt de la cascade, Plantation des deux arbres,* Rapport du jardinier en chef, République Française, Préfecture de la Seine, Direction administrative des Travaux de Paris, Service des Promenades, 24 April 1896.

50. Ibid.

51. Alphand, *Les Promenades de Paris,* lix.

52. Ibid., appendix, "Les Promenades Intérieurs de Paris—Planches"; William Robinson, *The Parks, Promenades, and Gardens of Paris, Described and Considered in Relation to the Wants of Our Own Cities, and the Public and Private Gardens* (London: J. Murray, 1869), 114, 126 illus.

53. AP, 1304W 210 Dossier 8, *Square Montholon; Grilles latérales,* Rapport du conducteur, Département de la Seine, Ville de Paris, Ponts et Chaussées, Service municipal des Travaux Publics, 24 April 1869.

54. Ibid.

55. AP, 1304W 114, *Parc des Buttes Chaumont: Voiture de chèvres pourvues d'un systeme de tramways à voie sans fin; Sieur Ader, Arrêté,* Préfecture du Département de la Seine, 2 October 1876.

56. AP, 1304W 114, *Parc des Buttes Chaumont: Demande de concession de Chalet Bébé; M. Bouquet,* Rapport de l'Inspecteur Caillas, République Française, Ponts et Chaussées, Direction des Travaux de Paris, Services de la Voie publique et des Promenades, 1889.

57. AP, 1304W 114, *Parc des Buttes Chaumont: Construction d'un Théâtre des marionnettes; Demande de M. Gautard,* Rapport de l'inspecteur, signé Lion, République Française, Pont et Chaussées,

Direction des Travaux de Paris, Service de la Voie publique et des Promenades, 22 April 1892; AP, 1304W 114, *Parc des Buttes Chaumont: Demande d'installation balançoires hygiéniques,* M. Gautard, Rapport de l'inspecteur, République Française, Préfecture de la Seine, Direction administrative des Travaux de Paris, Voirie et Travaux d'Ingénieurs, Service technique de l'Éclairage des Promenades et des Concessions, 2 May 1893; and AP, 1304W 114, *Parc des Buttes Chaumont: Ménage de chevaux en bois,* M. Ribert, *Arrêté,* République Française, Préfecture de la Seine, Direction Administrative des Travaux de Paris, Voirie et Travaux d'Ingénieurs, Service Technique de l'Éclairage des Promenades et des Concessions, 4 March 1893.

58. AP, 1304W 114, *Parc des Buttes Chaumont: Construction d'un Théâtre des marionnettes; Demande de M. Gautard,* 22 April 1892.

59. AP, Perotin 10653 106, *Bulletin quotidien de 22 septembre 1859,* M. Roche, Brigadier des Gardes, Département de la Seine, Ville de Paris, Pont et Chaussées, Service municipal, Promenades et Plantations.

60. AP, Perotin 10653 80, *Rapport du Garde, Square Monge,* Département de la Seine, Direction des Travaux de Paris, Promenades intérieures, 16 October 1874.

61. AP, VM90 2, *Square de la Mairie du 12e arrondissement, Établissement d'une bande de gazon autour du socle supportant le groupe en bronze,* Rapport du jardinier principal, République Française, Département de la Seine, Ville de Paris, Direction administrative des Service d'Architecture et des Promenades et Plantations, Service technique des Promenades et Plantations, 5 July 1900; *Square de la Mairie du XIIe, Établissement d'une bande de gazon autour du groupe de M. Hiolle,* Rapport du conservateur, République Française, Département de la Seine, Ville de Paris, Direction Administrative des Service d'Architecture et des Promenades et Plantations, 20 August 1900.

62. Alfred Delvau, *Les plaisirs de Paris: Guide pratique et illustré* (Paris: A. Faure, 1867), 42; Thomas Bullfinch, *Bullfinch's Mythology: The Age of Fable; the Age of Chivalry; Legends of Charlemagne* (London: Spring Books, 1964), 149–50.

63. Henri Dabot, *Souvenirs et impressions d'un bourgeois de quartier Latin de mai 1854 à mai 1869* (Péronne: Impr. E. Quentin, 1899), 128–29.

64. Ibid.

65. Archives de la Préfecture de la Police (APPP), CB.76.35, *Quartiers de Paris* (Combat) #296.

66. Ibid., #422.

67. AP, VM90 5, Letter from M. A. Pillot to M. le Préfet de la Seine, 15 June 1891 (regarding a broken arm); AP, VM90 5, Rapport du garde, Service de la Place des Vosges, République Française, Département de la Seine, Ville de Paris, Direction administrative des Service d'Architecture et des Promenades et Plantations, Service des Promenades; 15 July 1901 (exhibitionist); AP, Perotin 10653 81, Rapport des Gardes Roussel et Giampiétri, Service: Arts et Métiers, République Française, Département de la Seine, Ville de Paris, Direction Administrative des Service d'Architecture et des Promenades et Plantations, Service des Promenades, 13 June 1901 (lost child); and AP, 2380 40, Rapport du Garde Millet, Square de la Chapelle, République Française, Direction des Travaux de Paris, Promenades intérieurs, 21 August 1889 (bullying).

68. AP, VM90 5, Rapport du Garde Gras, Service de la Place des Vosges, République Française, Département de la Seine, Ville de Paris, Direction Administrative des Service d'Architecture et des Promenades et Plantations, Service des Promenades, Parc de Montsouris, 20 July 1900.

69. Ibid.

70. The register in which young Brunneau's trip to the local police station ought to have been recorded, APPP CB.14.45, was missing at the time of this research.

71. AP, 1304W 174, *Puits Artésien, Square Lamartine, Plainte de M. Lhuillier,* Rapport du jardinier en chef, République Française, Ponts et Chaussées, Direction des Travaux de Paris, Service des Promenades, 7 March 1888.

72. AP, 1304W 177, *Objet: Demande d'une gratification de 30f au profit des Sieurs Kohler, Vaqui* [sic], *et Bacourt auteurs de l'arrestation de Sieur Tisset surpris dans le Square des Innocents,* Letter from the Garde Général to M. l'Ingénieur en chef, 9 May 1862.

73. Ibid.

74. See also AP, Perotin 10653 86, *État hebdomadaire de renseignements sur le service des parcs, squares et jardins,* Département de la Seine, Ville de Paris, Pont et Chaussées Service municipal des Travaux Publics des Paris, 26 April 1890; AP, Perotin 10653 100, *Note pour M. le Directeur* (concerning theft of plant at Square Montmartre), Conservateur Flatraud, République Française, Département de la Seine, Ville de Paris, Direction administrative des Service d'Architecture et des Promenades et Plantations, Service des Promenades, Parc de Montsouris, 7 May 1902; AP, Perotin 10653 100, *Note pour M. le Directeur* (concerning theft of plants at Champs-Élysées), Conservateur Flatraud, République Française, Département de la Seine, Ville de Paris, Direction administrative des Service d'Architecture et des Promenades et Plantations, Service des Promenades, Parc de Montsouris, 17 May 1902; AP, Perotin 10653 100, *Champs-Élysées et* [illegible] *du Bois, Vols des plantes,* Letter from Le Préfet de la Seine to M. le Préfet de Police, 28 May 1902.

75. AP, VONC 908, *Ordre du Jour,* 1 February 1853, in *Registre du Garde Aubert,* garde portier du Bois de Boulogne, 1853–56.

76. Ibid.

77. AP, VONC 908, *Ordre du Jour,* 22 October 1855, in *Registre du Garde Aubert.* Pissot threatened the guards that he would withhold promotions, not that they would be fired. This gives some insight into the nature of employment as a park guard. It constituted a form of working retirement for former military, and they were sometimes quite lax in the execution of their duties.

78. AP, VONC 995, *Registre des procés verbaux des délits dans le Bois de Boulogne,* 1853–80. See also AP, VONC 881, *Registre de Garde Labadie,* garde forestier à la résidence Longchamps, 1861–63; and AP, VONC 1067, *Registre de Garde Bize,* chargé de la surveillance au [*illegible*] du lac supérieur, 1856–60.

79. AP, Perotin 10653 101, *État hebdomadaire de renseignements sur le service du Bois de Vincennes,* Département de la Seine, Ville de Paris, Pont et Chaussées Service municipal des Travaux Publics des Paris, 1864 (incomplete); AP, Perotin 10653 108 *État hebdomadaire de renseignements sur le service du Bois de Boulogne,* 1855, 1859, 1860 (incomplete), Département de la Seine, Ville de Paris, Pont et Chaussées, Service municipal des Travaux Publics de Paris, 1864.

80. AP, 2380W 20, *Park Montsouris, Passage reliant les gares de la Glacière et Sceaux-ceinture. Éclairage de nuit,* Rapport des l'ingénieurs ordinaire, signé Seilheimer, République Française, Ponts et Chaussées, Direction des Travaux de Paris, Service des Promenades, 12 October 1882; AP, 1304W 118, *Parc des Buttes Chaumont,* Rapport du sous-ingénieur, République Française, Ponts et Chaussées, Direction des Travaux de Paris, Service des Promenades, 18 April 1883.

81. AP, 2380 40, *Square de la Chapelle, Ordures disposées dans le square pendant la nuit,* République Française, Ponts et Chaussées, Direction des Travaux de Paris, Service des Promenades, 31 March 1886.

82. Ibid.

83. AP, 2380 40, *Square de la Chapelle, Ordures disposées dans le square pendant la nuit, Demande d'établissement de cabinets d'aisances,* République Française, Ponts et Chaussées, Direction des Travaux de Paris, Service des Promenades, 9 April 1886.

84. VM90 5, *Square Montholon, Plainte au sujet de la population peu recommandable qui accapare les bancs, M. Blondel,* Rapport du conservateur, République Française, Département de la Seine, Ville de Paris, Direction administrative des Service d'Architecture et des Promenades et Plantations, Service technique des Promenades, 6 September 1900.

85. Ibid.

86. Ibid.

87. AP, VONC 908, *Registre du Garde Aubert* (emphasis added).

88. Alain Corbin, *Women for Hire: Prostitution and Sexuality in France After 1850* (Cambridge, Mass: Harvard University Press, 1990), chap. 1.

89. Ibid., 9–11.

90. Ibid., 133–35.

91. AP, 2380 40, *Square de la Chapelle; Plainte de la façon s'est faite la police du square, Lettre sans adresse de Mme. Amélie Larcher,* Rapport de l'inspecteur, République Française, Préfecture de la Seine, Direction administrative des Travaux de Paris, Voirie et Travaux d'Ingénieurs, Service technique des Promenades, 22 July 1896.

92. Ibid.

93. Corbin, *Women for Hire,* 117–18.

94. Ibid., 119.

CHAPTER FIVE

Epigraph: Alfred Delvau, *Les plaisirs de Paris: Guide pratique et illustré* (Paris: A. Faure, 1867), 44.

1. See for example Stephen F. Eisenman, "Seeing Seurat Politically," *Art Institute of Chicago Museum Studies* 14, no. 2, The Grande Jatte at 100 (1989): 210–21, 247–49; and Susan Nochlin's article, "Seurat's Grande Jatte: An Anti-Utopian Allegory," in the same commemorative issue, 132–53, 241–42. Also see John Gage, "The Technique of Seurat: A Reappraisal," *The Art Bulletin* 69, no. 3 (September 1987): 448–54.

2. For more detail on the renovation of the Bois de Boulogne see Adolphe Alphand, *Les promenades de Paris: Histoire—description des embellissements—dépenses de création et d'entretien des Bois de Boulogne et de Vincennes, Champs-Élysées—Parcs—squares—boulevards—places plantées, Étude sur l'art des jardins et arboretum, 487 gravures sur bois, 80 sur acier, 23 chromolithographies* (1867; rpt. Princeton, N.J.: Princeton Architectural Press, 1984), 1–148; Georges-Eugène Haussmann, *Mémoires du Baron Haussmann 1: Grands Travaux de Paris* (Paris: Guy Durier, 1979), 183–209;

David H. Pinkney, *Napoléon III and the Rebuilding of Paris* (Princeton, N.J., Princeton University Press, 1958), 94–99; and Michel Carmona, *Haussmann* (Paris: Fayard, 2000), 240–42.

3. John W. Forney, *Letters from Europe* (Philadelphia: T. B. Peterson and Brothers, 1867), 184.

4. See Michael Conan, "The Coming of Age of the Bourgeois Garden," in John Dixon Hunt and Michael Conan, eds., *Tradition and Innovation in French Garden Art: Chapters of a New History* (Philadelphia: University of Pennsylvania Press, 2002), 160–83.

5. Delvau, *Les plaisirs de Paris*, 35.

6. Forney, *Letters from Europe*, 184 (emphasis added).

7. Mary E. Bouligny, *Bubbles and Ballast: Being a Description of Life in Paris during the Brilliant Days of the Empire: A Tour through Belgium and Holland, and a Sojourn in London* (Baltimore: Kelly, Piet, 1871), 51.

8. Ibid.

9. Alphand, *Les promenades de Paris*, 96.

10. Frédéric Loliée, *La fête impériale* (Paris: Jules Tallandier, 1912), 63.

11. Alphand, *Les promenades de Paris*, 97; Charles D. Warner, *Saunterings* (Boston: James R. Osgood and Co., 1872), 12.

12. Maurice Allem, *La vie quotidienne sous le Second Empire* (Paris: Librairie Hachette, 1948), 181.

13. Hervé Maneglier, *Paris impérial: La vie quotidienne sous le Second Empire* (Paris: Armand Colin, 1990), 160.

14. Amédée Achard, "Le Bois de Boulogne, les Champs-Élysées, le Bois et le Château de Vincennes," in *Paris guide par les principaux écrivains et artistes de la France, Deuxième partie: La vie* (Paris: A. Lacroix, 1867), 1237. In the nineteenth century, *grisette* referred to a young seamstress or working girl who cared for and was supported by her student lover. Grisettes were often abandoned when the bourgeois or aristocratic young men left the university. Nineteenth-century novelists and poets romanticized the figure of the grisette.

15. Édouard Gourdon, *Le Bois de Boulogne: Histoire, types, moeurs* (Paris: Librairie Charpentier, 1854), 183–86.

16. Ibid.

17. Haussmann, *Mémoires* 1: 204.

18. Alphand, *Les promenades de Paris*, 99.

19. Ibid.

20. Ibid., 100.

21. Ibid., 101.

22. Arsène Houssaye, *Les confessions: Souvenirs d'un demi-siècle, 1830–1880* (1885), as quoted in Maneglier, *Paris impérial*, 158.

23. "Le Club des patineurs au bois de Boulogne," *Le Monde illustré* 16, no. 404 (7 January 1865): 7.

24. AP, Perotin 10653 54, *Patinage*, Rapport signé Pissot, Département de la Seine, Ville de Paris, Direction d'inspection du Bois de Boulogne, 23 November 1855.

25. "Accident sur le lac du bois de Boulogne," *Le Monde illustré* 10, no 251 (1 February 1862): 70.

26. Ibid.

27. Henri Dabot, *Souvenirs et impressions d'un bourgeois de quartier Latin de mai 1854 à mai 1869* (Péronne: Impr. E. Quentin, 1899), 99–100.

28. Ibid.

29. AP, Perotin 10653 74, Dossier A, *Patinage,* Rapport du conservateur, Département de la Seine, Ville de Paris, Ponts et Chaussées, Service municipal des Travaux Publics de Paris, Service des Promenades et Plantations, 14 February 1862.

30. AP, VONC 881, *Registre de Garde Labadie,* garde forestier à la résidence Longchamps, 1861-63.

31. AP, Perotin 10653 74, Dossier A, Rapport du conservateur du Bois de Boulogne, Département de la Seine, Ville de Paris, Ponts et Chaussées, Service Municipal des Travaux Publics de Paris, Service des Promenades et Plantations, 20 July 1862.

32. AP, Perotin 10653 74, *Avis au public sur le Patinage,* Préfecture du Département de la Seine, Ville de Paris, Bois de Boulogne, November 1862.

33. AP, Perotin 10653 74, watercolor plan with the inscription, "vu et approuvé par l'Empereur le 27 janvier 1863."

34. AP, Perotin 10653 74, Dossier A, Rapport du conservateur du Bois de Boulogne, Département de la Seine, Ville de Paris, Ponts et Chaussées, Service municipal des Travaux Publics de Paris, Service des Promenades et Plantations, 20 July 1862.

35. Ibid.

36. AP, Perotin 10653 74, *Demande de louer des patins,* Rapport du conservateur du Bois de Boulogne, Département de la Seine, Ville de Paris, Ponts et Chaussées, Service municipal des Travaux Publics de Paris, Service des Promenades et Plantations, 19 November 1862.

37. Alphand, *Les promenades de Paris,* lviii-vix; Adolphe Alphand and Alfred Auguste Ernouf, *L'art des jardins. Parcs—jardins—promenades—étude historique—principles de la composition des jardins—plantations—décoration pittoresque et artistique des parcs et jardins publics; traité pratique et didactique* (Paris: J. Rothschild, 1886), 325-26, 348, 352.

38. "Le Patin," *Le Monde illustré* 10, no. 247 (4 January 1862): 7.

39. "Le Club des patineurs au bois de Boulogne," *Le Monde illustré* 16, no. 404 (7 January 1865): 7.

40. Ibid. It is unclear here if the author is using the word "races" to mean *classes*. The working class was sometimes referred to directly or obliquely in racial terms by reformers like Michel Lévy, for example in his *Traité d'hygiène publique et privée* (Paris: J.-B. Baillière, 1844). Or perhaps the term is used to mean *nationalities* in the nineteenth-century sense, as in "the French race." In either case, the author is reporting on the considerable diversity of the skating public.

41. Bouligny, *Bubbles and Ballast,* 171.

42. Arrêté préfectorial du 20 novembre 1871, in *Lois et règlements concernant les Bois de Boulogne et de Vincennes et les Promenades intérieures de la ville de Paris,* Préfecture du Département de la Seine, Ville de Paris, Direction administrative des Services d'Architecture et des Promenades et Plantations, 1900.

43. AP, Perotin 10653 46. This carton contains original identification badges issued by the Service des Promenades et Plantations and signed by Conservateur Pissot, as well as letters of application, letters of recommendation, and letters attesting to residence often signed at the local police station in the presence of an officer.

44. AP, VONC 881, *Registre de Garde Labadie;* AP, Perotin 10653 101, *État hebdomadaire de renseignements sur le service du Bois de Vincennes,* Département de la Seine, Ville de Paris, Pont et Chaussées, Service municipal des Travaux Publics de Paris, 1864 (incomplete).

45. AP, Perotin 10653 107, *Projets de règlements du bois de Boulogne*, proposé par l'ingénieur en Chef des Promenades et Plantations, September 1856.

46. AP, Perotin 10653 107, *Bois de Boulogne, Engins de pêche, Arrêté*, Préfecture du Département de la Seine, 2 October 1876.

47. AP, 1304W 113, *Bois de Boulogne et de Vincennes, Pêche à la linge*, Extrait du registre de procès verbaux des séances du Conseil municipal de la Ville de Paris, Préfecture de la Département de la Seine, Direction des Travaux de Paris, Séance du 13 mars 1872.

48. AP, 1304W 113, *Parc des Buttes Chaumont. Demande d'un permis de pêche, Sieur Pierre Charles Dubois,* Rapport du conducteur principal, Service technique des Promenades, 21 June 1876.

49. The two years with the fewest licenses were 1879 with thirteen and 1880 with five. All other years had twenty-one or more. The number of licenses granted each year and the cost, as well as the names, residence, and relationship of the licensees is contained in AP, 1304W 113 in a series of yearly licensing lists (1876–92) compiled by Ponts et Chaussées, Direction des Travaux de Paris, titled *État des sommes verser dans la caisse de la Ville de Paris pour permissions de pêche dans le lac du parc des Buttes Chaumont* (sometimes titled *État des sommes recouvrer par la Ville de Paris pour permissions de pêche dans le lac du parc des Buttes Chaumont*).

50. AP, 1304W 113, Newspaper clipping with notation, "*Bulletin Municipal* du 21 juin 1898, page 1660, 3e colonne."

51. AP, 1304W 114, *Parc de Montsouris; Demande d'autorisation de pêcher dans le lac, M. Kahn,* Rapport de l'Inspecteur Caillas. République Française, Ponts et Chaussées, Direction des Travaux de Paris, Services de la Voie publique et des Promenades, 1 July 1892.

52. Ibid.

53. AP, 2380 20, *Parc de Montsouris, Demande d'achat de poissons,* M. Travet, Rapport de l'inspecteur, République Française, Ponts et Chaussées, Direction des Travaux de Paris, Service des Promenades, 8 August 1890.

54. AP, 2380 20, *Parc de Montsouris; Pétition pour la délivrance de permis de pêche à la ligne dans le lac,* Rapport du jardinier du conducteur, République Française, Préfecture de la Seine, Direction administrative des Travaux de Paris, Voirie et Travaux d'Ingénieurs, Service technique des Promenades, 9 April 1897.

55. AP, Perotin 10653 74, Lettre à son excellence M. le Baron Haussmann, Sénateur, Préfet de la Seine de Thomas Sumpter, 18 August 1863.

56. AP, Perotin 10653 74, *Demande d'être autorisé à vendre aux membres du Cricket-Club;* Rapport du conservateur du Bois de Boulogne, Département de la Seine, Ville de Paris, Ponts et Chaussées, Service municipal des Travaux Publics, Service des Promenades et Plantations, 4 October 1863.

57. Ibid.

58. Ibid. *Guinguettes* were crude, country taverns, set up just outside the city's defensive walls. Since they were outside of the customs wall, the wine and spirits sold there were much cheaper than in Paris. Guinguettes became associated with shady dealing, smugglers, prostitutes, and pimps.

59. AP, Perotin 10653 74, *Demande d'être autorisé à vendre aux membres du Cricket-Club;* Rapport du Conservateur du Bois de Boulogne, Département de la Seine, Ville de Paris, Ponts et Chaussées Service municipal des Travaux Publics, Service des Promenades et Plantations, 4 October 1863.

60. AP, 10653 74, *Note pour M. l'Ingénieur en chef des promenades*, signée Alphand, Préfecture du Département de la Seine, Direction des Travaux de Paris, 6 May 1878.

61. Ibid.

62. Ibid.

63. Ibid. These same issues—damage to the park, impeding visitors from enjoying the park, and reasonable use of the space for games—appear again in an analogous response to a cricket team from the École Monge. The language in that report, also signed by Alphand, is remarkably similar, demonstrating the kind of continuity that characterized his decisions (AP, 10653 74, *Pelouses de Bois de Boulogne, Jeux de cricket et de ballon, Demande d'autorisation. Copie pour M. Bartet d'une lettre adressé à M. Périsse, rue de Rome 77*, signée Alphand, République Française, Ponts et Chaussées, Direction de Travaux de Paris, 25 January 1883).

64. AP, 1304W 114, *Parc des Buttes Chaumont, Autorisation demandé par M. LaBrousse pour y établir des régates et des course vélocipèdes*, Rapport de l'ingénieur en chef, Département de la Seine, Ville de Paris, Ponts et Chaussées, Direction de la Voie publique et des Promenades, Service des Promenades, 21 April 1868.

65. Ibid.

66. Ibid.

67. AP, VONC 69, *Règlements du Bois de Boulogne 10 October 1871*, Préfecture du Département de la Seine, Ville de Paris, Direction administrative des Services d'Architecture et des Promenades et Plantations, 1871; AP, VONC 69, *Règlements du Bois de Vincennes 10 October 1871*, Préfecture du Département de la Seine, Ville de Paris, Direction Administrative des Services d'Architecture et des Promenades et Plantations, 1871; and AP, VONC 69, *Règlements des Promenades intérieures de la ville de Paris*, Préfecture du Département de la Seine, Ville de Paris, Direction Administrative des Services d'Architecture et des Promenades et Plantations, 1871.

68. AP, 1304W 114, *Parc des Buttes Chaumont, Proposition tendant à interdire la circulation de vélocipèdes*, Rapport de l'inspecteur, République Française, Ponts et Chaussées, Direction des Travaux de Paris, Service de l'Inspection des Promenades, 3 Juin 1890.

69. Ibid.

70. AP, 1304W 114, *Parc des Buttes Chaumont, Circulation de vélocipèdes, Pétition de M. Richard*, Rapport de l'inspecteur signé Lion, République Française, Ponts et Chaussées, Direction des Travaux de Paris, Service de l'Inspection des Promenades, 23 September 1890.

71. Ibid.

72. Ibid. Lion's use of the plural here suggests that there may have been one or more accidents that followed the incident with the Briens boy.

73. AP, 1304W 114, Letter from M. Leotard to M. l'Ingénieur en chef, 11 November 1892; AP, 1304W 114, *Park des Buttes Chaumont. Autorisation de faire un course de Vélocipèdes, Club Vélocipèdique «Trois Étoiles»*, Rapport de l'inspecteur, République Française, Préfecture de la Seine, Direction administrative des Travaux de Paris, Service d'Éclairage, des Promenades et des Concessions, Inspection des Promenades, 30 November 1892.

74. AP, VONC 69, *Règlementation de la circulation de vélocipèdes dans les Bois de Boulogne et de Vincennes et de Parc Monceau*, République Française, Préfecture du Département de la Seine, Ville de Paris, 24 July 1896.

75. Stephen L. Harp, *Marketing Michelin: Advertising and Cultural Identity in Twentieth-Century France* (Baltimore: Johns Hopkins University Press, 2001), 55.

76. AP, 1304W 114, *Parc des Buttes Chaumont, Demande d'autorisation de laisser circuler les vélocipèdes sur la route reliant les rues Fessart et Secrétan, M. le Président du Touring-Club,* Rapport de l'inspecteur, République Française, Préfecture de la Seine, Direction administrative de la Voie publique et des Eaux et Égouts, Service des Promenades et Plantations, 22 November 1897.

77. AP, 1304W 114, *Règlementation de la circulation de vélocipèdes dans le Parc des Buttes Chaumont,* République Française, Préfecture du Département de la Seine, Ville de Paris, 24 January 1898.

CONCLUSION

1. AP, VM90 2, Letter from *M. le Secrétaire General de la Société de médecine publique et de génie sanitaire* to *M. le Ministre de l'Instruction publique,* 27 June 1912.

2. Ibid.

3. Adolphe Alphand, *Les promenades de Paris: histoire—description des embellissements—dépenses de création et d'entretien des Bois de Boulogne et de Vincennes, Champs-Élysées—Parcs—squares—boulevards—places plantées, Étude sur l'art des jardins et arboretum, 487 gravures sur bois, 80 sur acier, 23 chromolithographies* (1867; rpt. Princeton, N.J.: Princeton Architectural Press, 1984), lix.

4. Michel de Certeau, *The Practice of Everyday Life* (Berkeley: University of California Press, 1984), 96.

5. See Konstanze Sylvia Domhardt, "From 'Functional City' to 'Heart of the City': Green Space and Public Space in the CIAM Debates of 1942–1952," in *Greening the City: Urban Landscapes of the Twentieth Century,* ed. Dorothee Brantz and Sonja Dümpelmann (Charlottesville: University of Virginia Press, 2011).

6. Michael Bess, *The Light-Green Society: Ecology and Technological Modernity in France, 1960–2000* (Chicago: University of Chicago Press, 2003). See also Brantz and Dümpelmann, eds., *Greening the City,* part IV, and William Cronon, ed., *Uncommon Ground: Rethinking the Human Place in Nature* (New York: W. W. Norton & Co., 1996).

BIBLIOGRAPHY

•••• ••••

ARCHIVES DÉPARTMENTALES DE LA GIRONDE

2 O—Dossiers d'administration et communale.
    2 O 331-59 Bordeaux, Édifices et travaux, 1891-1914.

ARCHIVES NATIONALES

F1a—Objets généraux.
    F1a 604 Pétitions renvoyées par le Sénat, 1852-69.
F8—Police sanitaire.
    F8 210-213 Logements insalubres et habitations des ouvriers, 1850-98.
F14—Travaux Publics.
    F14 11459 Ponts et Chaussées, Dossiers personnels, Adolphe Alphand.

ARCHIVES DE PARIS

1304W	*Ville de Paris, Direction des Parcs, Jardins et Espaces Verts. 1870-1970 (approx.)* and contains notes, reports, minutes of meetings, work projects, maps, personnel documents, budgets, and a variety of memoranda and directives.
2380W	*Ville de Paris, Espaces Verts.*
D1U6	*Chambres correctionnelles: Jugements correctionels.*
D1Z	*Collection Lazare.*
D2Z	*Collection Blondel.*
D29Z 4	*Catalogue Exposition Universelle 1867.*

D9K13 170   *Rapports au conseil municipal Thermidor—1870.*
Perotin 10653   *Préfecture de la Seine. Services d'Architecture, des Promenades et Plantations, 1860–1930* Contains documents relating to the construction and management of municipal establishments.
TriBriand VM   *Architecture municipale 1852–1944.* There is a sub-series of 48 cartons, which deal with parcs, jardins, bois and squares, 1852–1942.
TriBriand VO   *Travaux Voie publique 1859–1942* Travaux d'aménagement de squares, jardins et concessions (chalets, urinoirs, voitures de place . . .); Circulation, signalisation, stationnement, fêtes foraines, etc. Despite chronology mostly expositions 1889 forward.
VK 395   *Funérailles de M. Adolphe Alphand.*
VM   *Bâtiments municipaux: construction et entretien 1808–1948.* The sub series VM90 contains 22 cotes relating to parks, gardens, woods and squares 1856–1943.
VONC   *Préfecture de la Seine, Service municipal des Travaux.* This collection spans the years 1810–1950 and contains documents pertaining to the general organization of the direction of public works, architecture, and promenades et plantations: such as budget, personnel, execution of projects.

ARCHIVES DE LA PRÉFECTURE DE POLICE DE PARIS

BB6   *Registre, Pédérastes et divers, 1875–1879.*
CB   *Quartiers de Paris, 1895–2000.*

NEWSPAPERS AND JOURNALS

Almanach du Magasin Pittoresque.
Bulletin du Ministère des travaux publics.
Le Correspondant.
Le Courrier municipal.
Figaro illustré.
Gazette municipal Revue municipale.
L'illustration journal universel.
Journal des villes et campagnes.
La Lanterne.
Le Monde illustré.
Le Moniteur universel.

The Morning News: The latest telegrams of the day.
La Nature.
L'opinion nationale.
Le Petit journal.
La Réforme du bâtiment.
Revue contemporaine.
Revue de deux mondes.
Revue d'hygiène thérapeutique.
Revue générale de l'architecture et des travaux publics.
La Revue municipal et gazette réunies.
La Rue: Paris pittoresque et populaire.
Le Siècle industriel.
The Spectator.
Le Temps.

PRIMARY SOURCES

Aguecheek. *My Unknown Chum.* Foreword by Henry Garrity. 1859. Rpt. New York: Devin-Adair Co., 1923.

Akerlio, M. le Docteur. *Les démolitions de Paris.* Paris: Chez tous les librairies, 1864.

Alphand, Adolphe. *Exposition universelle internationale de 1889 à Paris, palais, jardins, constructions diverses, installations générales: Monographie en collaboration avec ses chefs de service et avec le concours de M. Georges Berger.* Paris: J. Rothschild, 1892.

———. *Exposition universelle internationale de 1889. Direction générale des travaux. Rapport général sur la marche de la direction pendant l'année 1889.* Paris: Impr. nationale, 1890.

———. *Les promenades de Paris: Histoire—description des embellissements—dépenses de création et d'entretien des Bois de Boulogne et de Vincennes, Champs-Élysées—Parcs—squares—boulevards—places plantées. Étude sur l'art des jardins et arboretum, 487 gravures sur bois, 80 sur acier, 23 chromolithographies.* 1867. Rpt. Princeton, N.J.: Princeton Architectural Press, 1984.

———, and Alfred Auguste Ernouf. *L'art des jardins. Parcs—jardins—promenades—étude historique—principles de la composition des jardins—plantations—décoration pittoresque et artistique des parcs et jardins publics; traité pratique et didactique.* Paris: J. Rothschild, 1886.

André, Édouard. *L'art des jardins, Conférence faite le 26 août 1866 à Troyes dans la grande salle de l'Hôtel de Ville sous les auspices de la Société horticole, vigneronne, et forestière.* Troyes: Imprimerie et lithographie Dufour-Bouquot, 1866.

———. *L'art des jardins: Traité général de la composition des parcs et jardins.* 1879. Rpt. Marseille: Lafitte reprints, 1983.

———. *Discours prononcé par M. Édouard André, président de l'Union centrale (des beaux-arts appliqués à l'industrie) à la distribution des récompenses de l'exposition, le 14 décembre 1874.* Paris: Impr. de F. Debons, 1874.

Arnaud, Achille. *La pioche et le Luxembourg: Lettre d'un amateur de jardins aux Parisiens de la rive gauche.* Paris: Charlieu frères et Huillery, 1865.

Barillet-Deschamps, Jean-Pierre. *Création d'un jardin d'hiver au Long-Champs.* Bordeaux: Impr. des ouvriers associés, 1853.

Barthèlemy, Auguste. *Le vieux Paris et le nouveau dialogue en vers, par M. Barthèlemy.* Paris: Imp. Cordier, 1861.

Béjot, Eugène. *À Paris. Squares et jardins. Croquis lithographiques par Eugène Béjot.* Paris: [n.p.], 1896.

Bertelé, Alphonse-Balthazard. *Quelques mots sur les logements des ouvriers.* Strasbourg: Impr. d'Ad. Christophe, 1863.

Boué, Germaine. *Les Buttes-Chaumont: Notice Historique et Descriptive.* Paris: Chez tous les librairies, 1865.

———. *Les squares de Paris: La tour Saint-Jacques.* Paris: Librairie Centrale, 1864.

———. *Les squares de Paris: Notice sur le Parc de Monceaux* [sic]. Paris: Librairie Centrale, 1865.

———. *La tour du Temple.* Paris: Librairie centrale, 1864.

Bouligny, Mary E. *Bubbles and Ballast: Being a Description of Life in Paris during the Brilliant Days of the Empire: a Tour through Belgium and Holland, and a Sojourn in London.* Baltimore: Kelly, Piet, 1871.

Boussingault, Jean-Baptiste. *Agronomie, chimie agricole et physiologie, tome premier.* Paris: Mallet-Bachelier, 1860.

———. *Agronomie, chimie agricole et physiologie, tome deuxième.* Paris: Mallet-Bachelier, 1861.

Boyceau, Jacques. *Traité du jardinage selon les raisons de la nature et de l'art divisé en trois livres ensemble divers desseins de parterres, pelouzes, bosquets et autres ornemens servans* [sic] *à l'embellissement des jardins.* 1638. Rpt. Nördlingen: Verlag Dr. Alfons Uhl, 1997.

Bunel, H. *Établissements insalubres, incommodes, et dangereux; législation, inconvénients de ces établissements, et conditions d'autorisation ordinairement proposées par les conseils d'hygiène et de salubrité.* Paris: Berthoud, 1876.

Carrière, Élie-Abel. *Les arbres et la civilisation.* Paris: Chez l'auteur, 1857.

———. *Encyclopédie horticole.* Paris: La Maison Rustique, 1862.

———. *Jean-Pierre Barillet. Inauguration du monument élevé à sa mémoire.* Extrait de la *Revue Horticole,* 1 March 1876. Orlèans: Impr. De G. Jacob, 1876.

Cassell's Guide to Paris. Daily Itineraries. List of Hotels, Lodging houses, Restaurants etc. London: Cassell, Petter and Galpin, 1867.

Chadwick, Edwin. *Report to Her Majesty's Principal Secretary of State for the Home Office from the Poor Law Commissioners, On an Inquiry into the Sanitary Condition of the Labouring Population of Great Britain.* London: W. Clowes and Sons, 1842.

Champagnac, Jean-Baptiste-Joseph. *Sept jours à Paris. Promenades pittoresques et historiques par M. de Mirval. Nouvelle édition, revue et complétée par un professeur d'histoire.* Paris: E. Ducrocq, n.d.

Chancellor, E. Beresford. *The History of the Squares of London, Topographical and Historical.* Philadelphia: Lippincott, 1907.

Chevreul, Michel Eugène. *Recherches expérimentales sur la végétation par M. Georges Ville: Absorption de l'azote de l'air par les plantes.* Paris: Imprimerie de la Martinet, 1855.

Cheysson, Émile. *La question de habitations ouvrières en France et à l'étranger; la situation actuelle, ses dangers, ses remèdes. Conférence faite à l'Exposition d'hygiène de la caserne Lobau, le 17 juin 1886.* Paris: G. Masson, 1886.

Claude, Antoine. *Mémoires de Monsieur Claude, Chef de la Police de Sûreté sous le Second Empire, tome premier.* Paris: Jules Rouff, 1881.

Comte, Auguste, and Frederick Ferré. *Introduction to Positive Philosophy.* Indianapolis: Bobbs-Merrill, 1970.

Comte, Auguste, and H. S. Jones. *Early Political Writings.* Cambridge, U.K.: Cambridge University Press, 1998.

Comte, Auguste, and Gertrud Lenzer. *Auguste Comte and Positivism: The Essential Writings.* New York: Harper & Row, 1975.

Dabot, Henri. *Souvenirs et impressions d'un bourgeois de quartier Latin de mai 1854 à mai 1869.* Péronne: Impr. E. Quentin, 1899.

Daly, César, and Gabriel Jean Antoine Davioud. *Les théâtres de la place du Châtelet: Théâtre du Châtelet, Théâtre-lyrique.* Paris: Librairie générale de l'architecture et des travaux publics, 1866.

Darcel, Jean. *Exposition universelle de 1867 à Paris. Rapports du jury international publiés sous la direction de M. Michel Chevalier. Parcs et matériel de l'horticulture.* Paris: P. Dupont, 1867.

Delvau, Alfred. *Les plaisirs de Paris: Guide pratique et illustré.* Paris: A. Faure, 1867.

Descieux, Louis Cyprien. *Leçons d'hygiène à l'usage des enfants des écoles primaires.* Paris: Imprimerie et librairie classiques, 1858.

——— . *Manuel d'hygiène à l'usage des élèves des écoles normales primaires, des écoles spéciales, des lycées, collèges et séminaires.* Paris: P. Dupont, 1867.

Downing, A. J., and Henry Winthrop Sargent. *A Treatise on the Theory and Practice of Landscape Gardening, Adapted to North America: With a View to Improvement of Country Residences . . . ; with Remarks on Rural Architecture.* New York: A.O. Moore and Co, 1859.

Dubourg, G. *Ville de Paris. Promenades et plantations. Halles et marchés. Mémoire sur un projet de panneaux décoratifs pour affichage des actes, décrets et arrêtés préfectoraux et pour publicité permanente.* Paris: Impr. de A. Chaix, 1872.

Du Breuil, Alphonse. *Cours élémentaire théoretique et pratique d'arboriculture, comprenant l'étude des pépinières d'arbres et d'arbrisseaux forestiers, fruitiers et d'ornement; celle des plantations d'alignement forestières et d'ornement; la culture spéciale des arbres à fruits, à cidre et de ceux à fruits de table*. Paris: Langlois et Leclercq, 1853.

Du Camp, Maxime. *Paris ses organes, ses fonctions et sa vie dans la seconde moitié du XIXe siècle*. Vol. 6. 3rd ed. Paris: Librairie Hachette et cie., 1875.

Duvillers-Chasseloup, François Joseph. *Les parcs et jardins*. 1871.

Flachat, Stéphane. *Religion Saint-Simonienne, Le Choléra—Assainissement de Paris*. Paris: Everat Impr., 1832.

Fonssagrives, Jean-Baptiste. *L'éducation physique des garçons, ou, avis aux familles et aux instituteurs sur l'art de diriger leur santé et leur développement*. Paris: Ch. Delagrave et cie., libraires-éditeurs, 1870.

———. *L'éducation physique des jeunes filles; ou, avis aux mères sur l'art de diriger leur santé et leur développement*. Paris: L. Hachette, 1869.

———. *Hygiène et assainissement des villes*. Paris: J.-B. Baillière, 1874.

———. *La maison; étude d'hygiène et de bien-être domestiques*. Paris: Delagrave, 1871.

Forestier, Jean-Claude Nicholas. *Bagatelles et ses jardins*. Paris: Librairie horticole, 1910.

———. *Jardins, carnet de plans et de dessins*. Paris: Émile-Paul, frères, 1920.

Forney, John W. *Letters from Europe*. Philadelphia: T. B. Peterson and Brothers, 1867.

Frazee, Louis Jacob. *The Medical Student in Europe*. Maysville, Ky.: R. H. Collins, 1849.

Frégier, Henri-Antoine. *Des classes dangereuses de la population dans les grandes villes et des moyens de les rendre meilleures*. Paris: J.-B. Baillière, 1840.

Gautier, Hippolyte, and Adrien Desprez. *Les curiosités de l'Exposition de 1878: Guide du visiteur*. Paris: C. Delagrave, 1878.

Goodrich, Frank B. *Tricolored Sketches in Paris, During the Years 1851-2-3*. New York: Harper and Bros, 1855.

Gourdon, Édouard. *Le Bois de Boulogne: Histoire, types, moeurs*. Paris: Librairie Charpentier, 1854.

Halévy, Ludovic. *Carnets. Publiés avec une introduction et des notes par Daniel Halévy. Tome I, 1862–1869*. Paris: Calmann-Lévy, 1935.

Haussmann, Georges-Eugène. *Mémoires du Baron Haussmann*. 3 vols. Paris: Guy Durier, 1979.

Henrichs, P. *Napoleon III, président et empereur: Ou aperçu de ses principaux actes, écrits et discours, tels que les constatent* le Bulletin des lois *et le* Moniteur, *dans une première période de huit années de Décembre 1848 à 1857 Mars compris : indiquants, dans un ordre méthodique et chronologique, les lois, décrets et décisions tendus en faveur de l'armée de terre et de mer—de la classe ouvrière—du commerce—de l'industrie—de l'agriculture—des travaux publics—de la religion de l'enseignement—de nos colonies—et enfin dans un intérêt général : terminé par un coup d'oeil sur les décisions impériales pressés par sa Majesté l'Impératrice en faveur des classes laborieuses ou necessiteuses: Avec un analyse sommaire des principales dispositions de ces actes et des avantages qu'ils assurent*. Paris: Garnier frères, 1857.

Jeannel, Jean-François. "Des plantations d'arbres dans l'intérieur des villes au point de vue de l'hygiène publique, conférence faite au Jardin d'acclimatation, le 23 juillet 1872." *Bulletin mensuel de la Société d'acclimatation*, August 1872, 532–44.

Joanne, Adolphe. *Paris illustré en 1870 et 1876; guide de l'étranger et du Parisien contenant 442 vignettes dessinées sur bois, un plan de Paris et quartorze autres plans et un appendice pour 1876, Troisième édition.* Paris: Hachette, 1876.

———. *Sauvons le Luxembourg.* Paris: Sausset, 1866.

Joanne, Paul, ed. *Dictionnaire géographique et administratif de la France.* Paris: Librairie Hachette et cie., 1898.

John Murray (Firm). *Handbook for Visitors to Paris Containing a Description of the Most Remarkable Objects, in the City and Its Environs, with General Advice and Information for English Travellers in That Metropolis, and on the Way to It.* London: John Murray, 1878.

Jollivet, Gaston. *Souvenirs de la vie de plaisir sous le Second Empire.* Paris: Jules Tallandier, 1927.

Jouanet, Alexandre. *Mémoire sur les plantations de Paris.* Paris: Imprimerie Horticole de J.-B. Gros, 1855.

———. *Paris et ses plantations.* Paris: Typographie de Morris et cie., 1860.

Karl Baedeker (Firm). *Paris and Environs, With Routes from London to Paris; Handbook for Travellers.* Leipzig: K. Baedeker, 1907.

Lachaise, Claude. *Topographie médicale de Paris; ou, Examen général des causes qui peuvent avoir une influence marquée sur la santé des habitans [sic] de cette ville, le caractère de leurs maladies, et le choix des précautions hygiéniques qui leur sont applicables.* Paris: Baillière, 1822.

Lafenestre, Georges. *Notice sur la vie et les oeuvres de M. Alphand par M. Georges Lafenestre, membre de l'Académie des Beaux Arts, lue dans la séance du 29 juillet 1899.* Paris: Typographie de Firmin-Didot et cie., 1899.

Lazare, Felix, and Louis Lazare. *Dictionnaire administratif et historique des rues et monuments de Paris.* Paris: Au bureau de la revue municipale, 1855.

Lefebvre, Georges. *Service municipal. Plantations d'alignement, promenades, parcs et jardins publics.* Paris: P. Vicq-Dunod et cie., 1897.

Le Rousseau, Julien. *Conséquences du dégagement, de la limitation et de la réduction du jardin public dépendant du Luxembourg (décret du 25 novembre 1865).* Paris: Noirot, 1866.

Lespès, Léo. *Promenades dans Paris, par Léo Lespès (Timothée Trimm).* Paris: A. Faure, 1866.

Levallois, Jules. *Autour de Paris: promenades historiques.* 1883. Rpt. Nîmes: C. Lacour, 1992.

Lévy, Michel. *Traité d'hygiène publique et privée.* Paris: J.-B. Baillière, 1844.

Littré, Émile. *Auguste Comte et la philosophie positive.* Paris: L. Hachette et cie., 1864.

Lobet, J. *Le nouveau Bois de Boulogne et ses alentours; histoire, description et souvenirs.* Paris: Hachette, 1856.

Loliée, Frédéric. *La fête impériale.* Paris: Jules Tallandier, 1912.

Louis XIV, and Simon Hoog. *The Way to Present the Gardens of Versailles.* Éditions de la Réunion des Musées nationaux, 1992.

Madre, Ad. de. *Des ouvriers et des moyens d'améliorer leur condition dans les villes.* Paris: Hachette, 1863.

Martin, Alexis. *Tout autour de Paris: Promenades et excursions dans le Département de la Seine.* Paris: A. Hennuyer, 1890.

Massa, Philippe de. *Souvenirs et impressions: 1840–1871.* Paris: Calman Levy, 1897.

Masselin, Onésime, and Achille Juquin. *Description exacte des grands travaux du palais et du parc de l'Exposition universelle de 1867 à Paris.* Paris: O. Masselin, 1866.

Maw, Georges. *Exposition universelle française de 1867: Plan du palais de l'exposition : correspondance officielle et autre relative au plan promulgué par M. Le Play.* Londres: Impr. de Cox et Nyman, 1866.

McCabe, James D., Jr. *Paris by Sunlight and Gaslight: A Work Descriptive of the Mysteries and Miseries, the Virtues, the Vices, the Splendors and the Crimes of the City of Paris.* Philadelphia: National Publishing Co., 1869.

Morford, Henry. *Paris in 1867 or the Great Exposition: Its Sideshow and Excursions.* New York: Geo. W. Carleton and Co. Publishers, 1869.

Mouttet, Félix. *M. Haussmann et les Parisiens.* Paris: E. Dentu, 1868.

Murray, Eustace Clare Grenville. *High Life in France Under the Republic: Social and Satirical Sketches in Paris and the Provinces.* London: Vizetelly, 1884.

Napoleon III. *Napoleonic Ideas. Des idées napoléoniennes.* Ed. Brison Dowling Gooch. New York: Harper & Row, 1967

———. *The Political and Historical Works of Louis Napoleon Bonaparte, President of the French Republic. Now First Collected with an Original Memoir of His Life, Brought Down to the Promulgation of the Constitution of 1852; and Occasional Notes, Vols. 1 and 2.* New York: H. Fertig, 1972.

National Library of Medicine (U.S.). *Index-Catalogue of the Library of the Surgeon-General's Office.* Washington, D.C.: U.S. Dept. of Health, Education, and Welfare, Public Health Service [etc.] U.S. Govt. Printing Office, 1880.

———, and Carnegie Institution (Washington, D.C.). *Index Medicus: A Monthly Classified Record of the Current Medical Literature of the World.* Vol. 2. Washington, D.C.: U.S. Govt. Printing Office, 1880.

New York City. *PlaNYC: A Greener, Greater New York.* New York: City of New York, 2007.

North Peat, Anthony B. *Gossip from Paris during the Second Empire; Correspondence (1864–1869) of Anthony B. North Peat . . . Selected and arranged by A. R. Waller.* New York, D. Appleton and company, 1903.

Olmsted, Frederick Law. *The Papers of Frederick Law Olmsted.* Ed. Charles Capen McLaughlin and Charles E. Beveridge. 6 vols. Baltimore: Johns Hopkins University Press, 1977.

Parent-Duchâtelet, Alexandre-Jean-Baptiste. *De la prostitution dans la ville de Paris, considérée sous le rapport de l'hygiène publique, de la morale et de l'administration . . . précédé d'une notice historique sur la vie et les ouvrages de l'auteur, par Fr. Leuret.* Paris: J.-B. Baillière, 1836.

Paris guide par les principaux écrivains et artistes de la France. Deuxième partie: La vie. Paris: A. Lacroix, Verboeckhoven et cie., 1867.

Paris guide par les principaux écrivains et artistes de la France. Première partie: La science L'art. Paris: A. Lacroix, Verboeckhoven et cie., 1867.

Parville, Henri de. *Exposition universelle de 1867, Itinéraire dans Paris.* Paris: Garnier frères, 1867.

Pascal, Adrien. *Histoire de Napoléon III, empereur des Français; comprenant sa vie politique et privée, ses actes, ses discours, ses voyages, son avènement à l'empire, son mariage.* Paris: Barbier, 1853.

Paxton, Joseph, and John Lindley. *A Pocket Botanical Dictionary; Comprising the Names, History, and Culture of All Plants Known in Britain.* London: Bradbury and Evans, 1853.

Péan, Armand. *Parcs et jardins: Résumé des notes d'un praticien.* Paris: Leroux, 1878.

Persigny, Jean-Gilbert-Victor Fialin, and Henri de Laire. *Mémoires du duc de Persigny.* Paris: E. Plon, Nourrit & cie., 1896.

Le promeneur de Paris, 14 promenades dans les jardins parisiens. Le promeneur de Paris. Arles: Actes sud, 2003.

Proust, Marcel. *Remembrance of Things Past.* Trans. C. K. Scott Moncrieff. New York: Random House, 1934.

Pückler-Muskau, Hermann. *Puckler's Progress: The Adventures of Prince Pückler-Muskau in England, Wales, and Ireland as told in Letters to His Former Wife, 1826-9.* Trans. Flora Brennan. London: Collins, 1987

———. *A Regency Visitor: The English Tour of Prince Pückler-Muskau Described in His Letters, 1826-1828.* Ed. E. M. Butler. New York: Dutton, 1958.

———, and Eryck de Rubercy. *Aperçus sur l'art du jardin paysager. Assortis d'une petit revue de parcs anglais,* Collection L'esprit et les forms, 22. N.p.: Klincksieck, 1998.

Quatrefages, Armand de. *Exposition universelle des races canines au jardin zoologique d'acclimatation au Bois de Boulogne. Distribution des récompenses. Discours d'ouverture.* 1863.

Renaudin, Edmond. *Paris-Exposition; ou, Guide á Paris en 1867. Histoire, monuments, musées, théatres, curiosités, vie pratique avec la description du palais du Champ de Mars, et des environs de Paris. Orné de cartes, plans et gravures.* Paris: C. Delagrave, 1867.

Rengade, Jules. *Promenades d'un naturaliste aux environs de Paris, précédées d'une lettre à l'auteur, par M. Albert Millaud, et suivies d'un guide du naturaliste, de notes et de tableaux sur la flore et la faune parisiennes.* Paris: Du petit journal, 1866.

Repton, Humphry, and J. C. Loudon. *The Landscape Gardening and Landscape Architecture of the Late Humphrey Repton, Esq., Being His Entire Works on These Subjects.* London: Printed for the editor, and sold by Longman and Co. and A. C. Black, 1840.

Reynaud, Marius. *Les Buttes-Chaumont ou Saint-Chaumont. Les temps anciens et les temps modernes.* Châtellerault: Impr. de Bichon frères, 1870.

Rimmel, Eugene. *Recollections of the Paris exhibition of 1867.* Philadelphia: Lippincott, 1868.

Roberts, Henry. *The Dwellings of the Labouring Classes, Their Arrangement and Construction; Illustrated by a Reference to the Model Houses of the Society for Improving the Condition of the Labouring Classes, with Other Buildings Recently Erected: and an Appendix, Containing H.R.H. Prince Albert's Exhibition Model Houses, Hyde Park, 1851; the Model Cottages etc. Built by the Windsor Royal Society; with Plans and Elevations of Dwellings Adapted to Towns, as well as to Agricultural and Manufacturing Districts.* London: Seeleys, 1853.

Robinson, Charles Mulford. *The Improvement of Towns and Cities; or, The Practical Basis of Civic Aesthetics.* New York and London: G. P. Putnam and sons, 1901.

Robinson, William. *The Parks, Promenades, and Gardens of Paris Described and Considered in Relation to the Wants of Our Own Cities and of Public and Private Gardens.* London: J. Murray, 1869.

Rousseau, Jean-Jacques. *The Collected Writings of Rousseau.* Vol. 6: *Julie, or, The New Heloise: Letters of Two Lovers Who Live in a Small Town at the Foot of the Alps.* Trans. Philip Stewart and Jean Vaché. Hanover, N.H.: Dartmouth College, 1997.

Rouvet, Massillon. *Viollet-le-duc et Alphand au siège de Paris.* Paris: Librairies-Imprimeries Réunies, 1892.

Saint-Simon, Louis de Rouvroy. *Saint-Simon at Versailles.* Ed. Lucy Norton. New York: Harper, 1958.

Sala, Georges Augustus. *Notes and Sketches of the Paris Exhibition.* London: Tinsley Brothers, 1868.

Seine (France), and Benoiston de Châteauneuf. 1834. *Rapport sur la marche et les effets du choléra-morbus dans Paris et les communes rurales du département de la Seine par la commission nommée, avec l'approbation de m. le ministre du commerce et des travaux publics, par mm. les préfets de la Seine et de police. Année 1832.* Paris: Imprimerie royale, 1834.

Siebeck, R., Jules Rothschild, and Charles Naudin. *Guide pratique du jardinier paysagiste: Album de 24 plans coloriés sur la composition et l'ornementation des jardins d'agrément a l'usage des amateurs, propriétaires et architectes.* Paris: J. Rothschild, 1863.

Smith, Albert. *Paris and London; Humorous Sketches of Life in France and England.* London: F. Warne, 1867.

Taine, Hippolyte. *Notes on Paris.* Trans. with notes by John Austin Stevens. New York: H. Holt and Co, 1875.

———. *Voyage en Italie, tome 1, Naples et Rome.* Paris: Librairie Hachette, 1866.

Tardieu, Antoine Ambroise. *Dictionnaire d'hygiène publique et de salubrité, ou, Répertoire de toutes les questions relatives à la santé publique, considérées dans leurs rapports avec les subsistances, les épidémies, les professions, les établissements et institutions d'hygiène et de salubrité, complété par le texte des lois, décrets, arrêtés, ordonnances et instructions qui s'y rattachent.* Paris: Chez J. B. Baillière, 1852.

Tripp, Alonzo. *Crests from the Ocean-World; or, Experiences in a Voyage to Europe, Principally in France, Belgium, and England, in 1847 and 1848; Comprising Sketches in the Miniature Worlds, Paris, Brussels, and London. By a Traveller and Teacher.* Boston: Whittemore, Niles, and Hall, 1855.

Uzanne, Octave, Félix Vallotton, and François Courboin. *Les Rassemblements. Badauderies parisiennes. Physiologies de la rue.* Paris: Bibliophiles indépendants, 1896.

Vacherot, Jules. *Les parcs et jardins au commencement du XXe siècle. École française (Barillet-Deschamps).* Paris: O. Doin, 1908.

Vernois, Maxime. *Traité pratique d'hygiène industrielle et administrative: Comprenant l'étude des éstablissements insalubres, dangereux et incommodes.* Paris: Chez J.-B. Baillière, 1860.

Véron, Charles. *Les Buttes Chaumont; dédié à M. le Baron Haussmann, sénateur, préfet de la Seine.* Versailles: Impr. de E. Aubert, 1868.

Véron, Pierre. *Paris s'amuse.* Paris: E. Dentu, 1861.

Villermé, Louis-René. *De la mortalité dan les divers quartiers de la Ville de Paris.* N.p., 1830.

Warner, Charles D. *Saunterings.* Boston: James R. Osgood and Co., 1872.

X, Arthur, H. Legludic, and Jacques Chazaud. *Mémoires d'un travesti, prostitué, homosexuel: "La Comtesse" 1850–1861.* Paris: Harmattan, 2000.

Zola, Émile. *The Kill.* Trans. A. Teixeira de Mattos. London: Elek Books Ltd., 1969.

———. *Le Ventre de Paris.* Paris: Fasquelle, 1968.

SECONDARY SOURCES

Accampo, Elinor, Rachel Fuchs, and Mary Lynn Stewart. *Gender and the Politics of Social Reform in France, 1870–1914.* Baltimore: Johns Hopkins University Press, 1995.

Adams, William Howard. *The French Garden: 1500–1800.* World Landscape Art and Architecture Series. New York: Braziller, 1979.

Alexander, Christopher, Sara Ishikawa, and Murray Silverstein. *A Pattern Language: Towns, Buildings, Construction.* New York: Oxford University Press, 1977.

Allem, Maurice. *La vie quotidienne sous le Second Empire.* Paris: Librairie Hachette, 1948.

Aprile, Sylvie. *Le Deuxième République et le Second Empire: 1848–1870.* Paris: Éditions Pygmalion, 2000.

Arnold, Dana. *Rural Urbanism: London Landscapes in the Early Nineteenth Century.* Manchester, U.K.: Manchester University Press, 2005.

Aulie, Richard P. "Boussingault and the Nitrogen Cycle." *Proceedings of the American Philosophical Society* 114, no. 6 (December 1970): 435–79.

Baguley, David. *Napoleon III and His Regime: An Extravaganza.* Modernist studies. Baton Rouge: Louisiana State University Press, 2000.

Barles, Sabine, and Laurence Lestel. "The Nitrogen Question: Urbanization, Industrialization, and River Quality in Paris, 1830–1939." *Journal of Urban History* 33, no. 5 (July 2007): 794–812.

Beecher, Jonathan. *Victor Considerant and the Rise and Fall of French Romantic Socialism.* Berkeley: University of California Press, 2001.

Benjamin, Walter. *Charles Baudelaire: A Lyric Poet in the Era of High Capitalism.* Trans. from German by Harry Zohn. London: NLB, 1973.

Berlanstein, Lenard R. *Big Business and Industrial Conflict in Nineteenth-Century France: A Social History of the Parisian Gas Company.* Berkeley: University of California Press, 1991.

Bess, Michael. *The Light-Green Society: Ecology and Technological Modernity in France, 1960–2000.* Chicago: University of Chicago Press, 2003.

Blanning, T. C. W. *The Culture of Power and the Power of Culture, Old Regime Europe, 1660–1789.* Oxford, U.K.: Oxford University Press, 2002.

Bowie, Karen. *La modernité avant Haussmann: Formes de l'espace urbain à Paris, 1801–1853.* Paris: Recherches, 2001.

Brantz, Dorothee, and Sonja Dümpelmann, eds. *Greening the City: Urban Landscapes of the Twentieth Century.* Charlottesville: University of Virginia Press, 2011.

Briggs, Asa. *Victorian Cities.* Berkeley: University of California Press, 1993.

Bryant, Christopher. *Positivism in Social Theory and Research.* New York: St. Martin's Press, 1985.

Buchanan, Ian. *Michel de Certeau: Cultural Theorist.* London: SAGE Publishing, 2000.

Bullfinch, Thomas. *Bullfinch's Mythology: The Age of Fable; the Age of Chivalry; Legends of Charlemagne.* London: Spring Books, 1964.

Buse, Peter. *Benjamin's Arcades: An Unguided Tour.* Manchester, U.K.: Manchester University Press, 2006.

Carmona, Michel. *Haussmann.* Paris: Fayard, 2000.

Cars, Jean des and Pierre Pinon. *Paris-Haussmann: "Le Pari d'Haussmann."* Paris: Pavillon de l'Arsenal, 1991.

Certeau, Michel de. *The Practice of Everyday Life.* Berkeley: University of California Press, 1984.

Champigneulle, Bernard. *Promenades dans les Jardins de Paris, ses Bois et ses Squares.* Paris: Les Librairies Associés, 1965.

Chantal, Auré. *Paris XIXe–XXe siècles, urbanisme, architecture, espaces verts: Bibliographie et sources imprimés à la Bibliothèque des Archives de Paris.* Département de Paris: Archives de Paris, 1995.

Charlton, D. G. *Positivist Thought in France During the Second Empire, 1852–1870.* Oxford, U.K.: Clarendon Press, 1959.

Chevalier, Louis. *Laboring Classes and Dangerous Classes in Paris During the First Half of the Nineteenth Century.* New York: H. Fertig, 1973.

Christiansen, Rupert. *Tales of the New Babylon: Paris 1869-1875*. London: Sinclair-Stevenson, 1994.

Clark, Peter. *The European City and Green Space: London, Stockholm, Helsinki and St. Petersburg, 1850-2000*. Aldershot, England: Ashgate, 2006.

Clark, T. J. *The Painting of Modern Life: Paris in the Art of Manet and His Followers*. Princeton, N.J.: Princeton University Press, 1984.

Clozier, R. "Géographie et urbanisme: Le réseau d'assainissement de la région Parisienne et se conditions géographiques." *Urbanisme et architecture; études écrites et publiées en l'honneur de Pierre Lavedan*, April 1954, 81-86.

Cole, Joshua. *The Power of Large Numbers: Population, Politics, and Gender in Nineteenth-Century France*. Ithaca, N.Y.: Cornell University Press, 2000.

Coleman, William. *Death Is a Social Disease: Public Health and Political Economy in Early Industrial France*. Madison: University of Wisconsin Press, 1982.

Corbin, Alain. *Women for Hire: Prostitution and Sexuality in France after 1850*. Cambridge, Mass: Harvard University Press, 1990.

Cosgrove, Denis E., and Stephen Daniels. *The Iconography of Landscape: Essays on the Symbolic Representation, Design, and Use of Past Environments*. Cambridge, U.K.: Cambridge University Press, 1988.

Cronon, William, ed. *Uncommon Ground: Rethinking the Human Place in Nature*. New York: W. W. Norton & Co., 1996.

Duveau, Georges. *La vie ouvrière en France, sous le Second Empire*. Paris: Gallimard, 1946.

Eisenman, Stephen F. "Seeing Seurat Politically." *Art Institute of Chicago Museum Studies* 14, no. 2, The Grande Jatte at 100 (1989): 210-21, 247-49.

Erlanger, Philippe. *Louis XIV*. New York: Praeger Publishers, 1970.

Faure, Alain, Alain Dalotel, and Jean-Claude Freiermuth. *Aux origines de la Commune: Le mouvement des réunions publiques à Paris, 1868-1870*. Paris: F. Maspero, 1980.

Forestier, Jean-Claude Nicolas. *Bagatelle et ses jardins*. Paris: Librairie horticole, 1910.

———, Bénédicte Leclerc, and Salvador Tarragó. *Grandes villes et systèmes de parcs: suivi de deux mémoires sur les villes impériales du Maroc et sur Buenos Aires*. Paris: Éditions Norma, 1997.

Fuchs, Rachel. *Abandoned Children: Foundlings and Child Welfare in Nineteenth-Century France*. Albany: State University of New York Press, 1984.

———. *Contested Paternity: Constructing Families in Modern France*. Baltimore: Johns Hopkins University Press, 2008.

———. *Poor and Pregnant in Paris: Strategies for Survival in the Nineteenth Century*. New Brunswick, N.J.: Rutgers University Press, 1992.

Furet, François. *Revolutionary France, 1770-1880*. Trans. Antonia Nevill. Oxford, U.K.: Blackwell, 1992.

Gage, John. *Color and Culture: Practice and Meaning from Antiquity to Abstraction*. Berkeley: University of California Press, 1999.

———. "The Technique of Seurat: A Reappraisal." *The Art Bulletin* 69, no. 3 (September 1987): 448–54.

Goldstein, Claire. *Vaux and Versailles: The Appropriations, Erasures, and Accidents That Made Modern France.* Philadelphia: University of Pennsylvania Press, 2008.

Goodman, David C., and Colin Chant. *European Cities & Technology: Industrial to Post-Industrial City.* London: Routledge in association with Open University, 1999.

Gould, Roger V. *Insurgent Identities: Class, Community, and Protest in Paris from 1848 to the Commune.* Chicago: University of Chicago Press, 1995.

Green, Nicholas. *The Spectacle of Nature: Landscape and Bourgeois Culture in Nineteenth-Century France.* Manchester, U.K.: Manchester University Press, 1990.

Hadfield, Miles, Robert Harling, and Leonie Highton. *British Gardeners: A Biographical Dictionary.* London: A. Zwemmer Ltd., 1980.

Harp, Stephen L. *Marketing Michelin: Advertising and Cultural Identity in Twentieth-Century France.* Baltimore: Johns Hopkins University Press, 2001.

Harrison, Carol. *The Bourgeois Citizen in the Nineteenth-Century France: Gender, Sociability, and the Uses of Emulation.* Oxford, U.K.: Oxford University Press, 1999.

Harvey, David. *Paris, Capital of Modernity.* New York: Routledge, 2003.

Hayden, Dolores. *Building Suburbia: Green Fields and Urban Growth, 1820–2000.* New York: Pantheon Books, 2003.

Hazareesingh, Sudhir. *Intellectual Founders of the Republic: Five Studies in Nineteenth-Century French Republican Political Thought.* Oxford, U.K.: Oxford University Press, 2001.

Hazlehurst, F. Hamilton. *Gardens of Illusion: The Genius of André le Nostre* [sic]. Nashville: Vanderbilt University Press, 1980.

———. *Jacques Boyceau and the French Formal Garden.* Athens: University of Georgia Press, 1966.

Higonnet, Patrice L. R. *Paris: Capital of the World.* Cambridge, Mass.: Belknap Press of Harvard University Press, 2002.

Hillairet, Jacques. *Dictionnaire historique des rues de Paris.* Paris: Éditions de minuit, 1963.

Hopkins, Richard S. "*Sauvons le Luxembourg*: Urban Greenspace as Private Domain and Public Battleground, 1865–1867." *Journal of Urban History* 37, no. 1 (January 2011): 43–58.

Horne, Alistair. *Seven Ages of Paris.* New York: A. A. Knopf, 2002.

Hunt, John Dixon. *The Figure in the Landscape: Poetry, Painting, and Gardening during the Eighteenth Century.* Baltimore: Johns Hopkins University Press, 1976.

———. *Gardens and the Picturesque: Studies in the History of Landscape Architecture.* Boston: MIT Press, 1992.

———. *William Kent, Landscape Garden Designer: An Assessment and Catalogue of His Designs.* London: A. Zwemmer Ltd., 1987.

———, and Michel Conan, eds. *Tradition and Innovation in French Garden Art: Chapters of a New History.* Philadelphia: University of Pennsylvania Press, 2002.

Hunt, Lynn Avery. *The French Revolution and Human Rights: A Brief Documentary History*. Bedford series in history and culture. Boston: Bedford Books of St. Martin's Press, 1996.
Imbs, Paul, and B. Quemada. *Trésor de la langue française; dictionnaire de la langue du XIXe et du XXe siècle, 1789–1960*. Paris: Éditions du Centre national de la recherche scientifique, 1971.
Jackson, Julian. *Living in Arcadia: Homosexuality, Politics, and Morality in France from the Liberation to AIDS*. Chicago: University of Chicago Press, 2009.
Jacquemet, Gèrard. *Belleville au XIXe Siècle: Du faubourg à la ville*. Paris: École des Hautes Études en Sciences Sociales, 1984.
Jones, Colin. *Paris: Biography of a City*. New York: Viking, 2005.
Jordan, David. *Transforming Paris: The Life and Labors of Baron Haussmann*. New York: Free Press, 1995.
Kluckert, Ehrenfried. *European Garden Design from Classical Antiquity to the Present Day*. Ed. Rolf Toman. Cologne: Könemann, 2000.
Kudlick, Catherine. *Cholera in Post-Revolutionary Paris: A Cultural History*. Berkeley: University of California Press, 1996.
Langlois, Gilles-Antoine. *Folies, tivolis et attractions: Les premiers parcs loisirs parisiens*. Paris: Délégation à l'action artistique de la ville de Paris, 1991.
Lavedan, Pierre. *Histoire de l'urbanisme à Paris*. Nouvelle histoire de Paris. Paris: Association pour la publication d'une histoire de Paris, 1975.
———. "Projets de Napoléon 1er pour l'ouest de Paris." *La Vie urbaine* 59 (1951): 1–10.
Lawrence, Henry W. "The Greening of the Squares of London: Transformation of Urban Landscapes and Ideals." *Annals of the Association of American Geographers* 83, Issue 1 (March 1993): 90–118.
Léri, Jean-Marc. *La presse à Paris 1851–1881: Hôtel de Lamoignon, 15 avril–30 juin 1983, Mairie du IIe arrondissement, octobre-novembre 1983*. N.p., 1983.
Lévêque, Jean-Jacques. *Jardins de Paris*. Paris: Hachette, 1982.
Limido, Luisa. *L'art des jardins sous le Second Empire: Jean-Pierre Barillet-Deschamps, 1824–1873*. Seyssel: Champ Vallon, 2002.
Loyer, Francois, and Stan Neumann. *Paris, roman d'une ville*. Ho-Ho-Kus, N.J.: Roland Collection, 1990.
Mabberley, D. J. *The Plant-Book: A Portable Dictionary of the Higher Plants Utilising Cronquist's* An Integrated System of Classification of Flowering Plants *(1981) and Current Botanical Literature, Arranged Largely on the Principles of Editions 1–6 (1896/97–1931) of Willis's* A Dictionary of the Flowering Plants and Ferns. Cambridge, U.K.: Cambridge University Press, 1987.
Maccubbin, Robert P., and Peter Martin, eds. *British and American Gardens in the Eighteenth Century: Eighteen Illustrated Essays on Garden History*. Williamsburg, VA: Colonial Williamsburg Foundation, 1984.

Malet, Henri. *Le Baron Haussmann et la rénovation de Paris.* Paris: Les éditions municipales, 1973.

Maneglier, Hervé. *Paris impérial: La vie quotidienne sous le Second Empire.* Paris: Armand Colin, 1990.

Marcus, Sharon. *Apartment Stories: City and Home in Nineteenth-Century Paris and London.* Berkeley: University of California Press, 1999.

McMillan, James F. *Napoleon III.* London: Longman, 1991.

Meinig, D. W., and John Brinckerhoff Jackson. *The Interpretation of Ordinary Landscapes: Geographical Essays.* New York: Oxford University Press, 1979.

Miller, Michael Barry. *The Bon Marché: Bourgeois Culture and the Department Store, 1869–1920.* Princeton, N.J.: Princeton University Press, 1981.

Moncan, Patrice de. *Les jardins du baron Haussmann.* Paris: Éditions du Mécène, 1992.

———, and Claude Heurteux. *Le Paris d'Haussmann.* Paris: Mécène. 2002.

Mukerji, Chandra. *Territorial Ambitions and the Gardens of Versailles.* New York: Cambridge University Press, 1997.

Newton, Norman T. *Design on the Land: The Development of Landscape Architecture.* Cambridge, Mass.: Harvard University Press, 1971.

Nochlin, Susan. "Seurat's Grande Jatte: An Anti-Utopian Allegory." *Art Institute of Chicago Museum Studies* 14, no. 2, The Grande Jatte at 100 (1989): 132–53, 241–42.

Papayanis, Nicholas. *Planning Paris before Haussmann.* Baltimore: Johns Hopkins University Press, 2004.

Peniston, William A. *Pederasts and Others: Urban Culture and Sexual Identity in Nineteenth-Century Paris.* New York: Harrington Park Press, 2004.

Peterson, Jon A. "The City Beautiful Movement: Forgotten Origins and Lost Meanings." *Journal of Urban History,* 2: 4 (August 1976): 415–434.

Phillips, Peggy A. *Modern France: Theories and Realities of Urban Planning.* Lanham, MD: University Press of America, 1987.

Pickering, Mary. *Auguste Comte: An Intellectual Biography.* Cambridge, U.K.: Cambridge University Press, 1993.

Pinkney, David H. "Migrations to Paris during the Second Empire." *The Journal of Modern History* 25, no. 1 (March 1953): 1–12.

———. "Money and Politics in the Rebuilding of Paris, 1860–1870." *The Journal of Economic History* 17, no. 1 (March 1957): 45–61.

———. *Napoléon III and the Rebuilding of Paris.* Princeton, N.J.: Princeton University Press, 1958.

———. "Napoléon III's Transformation of Paris: The Origins and Development of the Idea." *The Journal of Modern History* 27, no. 2 (June 1955): 125–134.

Pinon, Pierre. *Atlas du Paris haussmannien: La ville en héritage du Second Empire à nos jours.* Paris: Parigramme, 2002.

———. *Paris, biographie d'une capitale.* Paris: Hazan, 1999.

Pitte, Jean-Robert. *Histoire du paysage français*. Paris: Tallandier, 1983.
Plessis, Alain. *The Rise and Fall of the Second Empire, 1852–1871*. Trans. Jonathan Mandelbaum. London: Cambridge University Press, 1985.
Prendergast, Christopher. *Paris and the Nineteenth Century*. Oxford, UK: Blackwell, 1992.
Richardson, Joanna. *La Vie Parisienne: 1852–1870*. New York: The Viking Press, 1971.
Rifkin, Adrian and Roger Thomas. *Voices of the People: The Social Life of 'La Sociale' at the End of the Second Empire*. London: Routledge and Kegan Paul, 1988.
Saalman, Howard. *Haussmann: Paris Transformed*. New York: G. Braziller, 1971.
Sablet, Michel de. *Des espaces urbains agréables à vivre: Places, rues, squares et jardins*. Paris: Éditions du Moniteur, 1988.
Schott, Dieter, Bill Luckin, and Geneviève Massard-Guilbaud. *Resources of the City: Contributions to an Environmental History of Modern Europe*. Aldershot, England: Ashgate, 2005.
Schultz, Stanley K. "To Engineer the Metropolis: Sewers, Sanitation, and City Planning in Late-Nineteenth-Century America." *The Journal of American History* 65, no.1 (September 1978): 389–411.
Schwartz, Vanessa R. *Spectacular Realities: Early Mass Culture in Fin-De-Siècle Paris*. Berkeley: University of California Press, 1998.
Scott, Joan Wallach. *Gender and the Politics of History*. New York: Columbia University Press, 1988.
Shapiro, Anne-Louise. "Housing Reform in Paris: Social Space and Social Control." *French Historical Studies* 12, no. 4 (Autumn 1982): 486–507.
Simon, W. M. *European Positivism in the Nineteenth Century: An Essay in Intellectual History*. Ithaca, N.Y.: Cornell University Press, 1963.
Smith, William H. C. *Napoléon III: The Pursuit of Prestige*. London: Collins and Brown, 1991.
——. *Second Empire and Commune: France, 1848–1871*. Seminar studies in history. London: Longman, 1996.
Spirn, Anne Whiston. *The Granite Garden: Urban Nature and Human Design*. New York: Basic Books, 1984.
Steenbergen, Clemens M., Wouter Reh, and Gerrit Smienk. *Architecture and Landscape: The Design Experiment of the Great European Gardens and Landscapes*. New York: Prestel, 1996.
Stromberg, Roland N. *European Intellectual History Since 1789*. Englewood Cliffs, N.J.: Prentice-Hall, 1990.
Taylor, Katherine Fischer. *In the Theater of Criminal Justice: The Palais de Justice in Second Empire Paris*. Princeton, N.J.: Princeton University Press, 1993.
Texier, Simon, and Béatrice de Andia. *Les parcs et jardins dans l'urbanisme parisien: XIXe–XXe siècles*. Collection Paris et son patrimoine. Paris: Action artistique de la ville de Paris, 2001.

Thompson, Victoria. "Telling 'Spatial Stories': Urban Space and Bourgeois Identity in Early Nineteenth-century Paris." *Journal of Modern History* 75, no. 1 (September 2003): 523–56.

———. *The Virtuous Marketplace: Women and Men, Money and Politics in Paris, 1830–1870.* Baltimore: Johns Hopkins University Press, 2000.

Tobey, George B. *A History of Landscape Architecture: the Relationship of People to Environment.* New York: American Elsevier Pub. Co, 1973.

Tombs, Robert. *The Paris Commune, 1871.* London: Longman, 1999.

Tuan, Yi-Fu. *Space and Place: The Perspective of Experience.* Minneapolis: University of Minnesota Press, 1977.

———. "Thought and Landscape." In *The Interpretation of Ordinary Landscapes,* ed. D. W. Meinig. Oxford, U.K.: Oxford University Press, 1979.

Turner, Victor. *Dramas, Fields, and Metaphors: Symbolic Action in Human Society.* Symbol, Myth, and Ritual Series. Ithaca, N.Y.: Cornell University Press, 1974.

———. *The Ritual Process: Structure and Anti-Structure.* Lewis Henry Morgan Lectures, 1966. Chicago: Aldine Pub. Co., 1969.

Watkin, David. *The English Vision: The Picturesque in Architecture, Landscape, and Garden Design.* Icon editions. New York: Harper and Row, 1982.

Wernick, Andrew. *Auguste Comte and the Religion of Humanity: The Post-Theistic Program of French Social Theory.* Cambridge, U.K.: Cambridge University Press, 2001.

Zeldin, Theodore. *The Political System of Napoleon III.* London: Macmillan, 1958.

INDEX

• • • • • • • •

access issues and petitions, 95–106
Achard, Amédée, 135
agency, human, 7–8
air, fresh (*salubrité*), 39, 42–43, 54–58
Alleziette, Charles d', 28
Alphand, Jean-Charles Adolphe:
    background of, 19, 76; in Bordeaux,
    174n44; children's interests and, 116;
    complaints and petitions to, 88, 106–7,
    112, 122; criticism of, 32–33; guards and,
    64–66; Haussmann compared to, 19–20;
    horticultural museum proposal and, 29;
    ice production and, 27; *jardin anglais*
    style and, 8; landscape design and, 153–54;
    New York City and, 156; nurseries and,
    21; park service and, 2, 18–19; philosophy
    of, 5, 154–55; private spaces and, 137; on
    public health, 53–54; on soils, 23–24;
    sports and leisure activities and, 132–33,
    135–36, 140, 141–42, 143, 147–48
André, Édouard, 29, 48, 155–56, 174n44
Anvers, Square d', 53
Arboussier, M. d', 83
Archevêché, Square de l', 117–18
Aubert, M., 125
Avenue de la République, Square (now Square
    Samuel de Champlain), 84

Bacourt, M., 122
Barillet-Deschamps, Jean-Pierre, 25–26, 29,
    155, 174n44
Baronnier, M., 111
Belgrand, Eugène, 18
Belleville, Square de, 53
Belotte, M., 84–85
Berger, Jean-Jacques, 15
Bertelé, Alphonse, 45–46
Bess, Michael, 158
bicycling, 148–51
Blin, Mme., 87
Blondel, M., 124–25
bois, defined, x
Bois de Boulogne: Alphand and, 18; bicycles
    in, 149; carriage and horse rentals in,
    78; Cercle des patineurs and ice skating,
    135–43; cricket in, 146–48; criticism of, 32;
    ducks in, 25–26; firewood gathering by
    indigents in, 122–23; fish in, 27; fishing in,
    143–44; grass in, 23–24; guards in, 63, 64–
    65; horse races at Longchamps, 134–35; ice
    production in, 27; Jouanet on, 17–18; lakes
    and *tour des lacs,* 132–34; Mare St. James,
    139; night patrols in, 123; prostitutes in, 125
Bois de Vincennes: bicycles in, 149; carriage
    and horse rentals in, 78; criticism of, 32;

Bois de Vincennes (*continued*):
ducks in, 27; fishing in, 144; guards in, 63, 64; homosexual activity and surveillance in, 177n26; ice production in, 27; ice skating in, 137; night patrols in, 123; prostitutes in, 125
Boisschier, M., 110
Bonaparte, Louis-Napoleon, 45, 46
Bordeaux, 30–32, 170n75, 174n44
Bordelais, Parc (Bordeaux), 30–32
Borel, M., 141
Bos, Charles, 105
Boué, Germaine, 17
Bouligny, Mary, 133–34, 142
Bouquet, M., 79, 82–83, 145
Boussingault, Jean-Baptiste, 48–49
Bréton, M., 122
Bridoux, M., 83–84
Brincourt, Mme., 68–69
British Reform movement, 157
Brunneau, Émile, 121
Buchet, M., 110, 111
Bühler, Eugène, 29, 31
Buttes Chaumont, Parc des: bicycles in, 148–51; carriage and horse rentals in, 78; child predator incidents in, 119–20; children in, 116–17; concessions in, 78–83, 109–10; fishing in, 144–45; guards in, 64, 65–66; horticultural museum proposal, 28–29; inauguration of, 33; location of, 53; as model for Bordeaux, 31; Napoleon III and, 183n14; petitions and negotiations on, 96–100, 98(map), 102–3, 105–6, 107; yacht racing proposal, 148–49

Caillas, Inspector, 117, 124
cantonniers, 64–65, 74–76, 110
Carmona, Michel, 53
carnivals (*fêtes foraines*), 106–7
carriage and horse rentals, 78
Cercle des patineurs, 135–43
Certeau, Michel de, 7, 156–57
Chapelle, Square de la, 53, 70–72, 124, 126–27

Chapelle Expiatoire (Square Louis XVI), 95–96, 97(map)
Chareille, Jean, 75
Chateau de la Muette, 21, 169n47
Chemin de Fer de Ceinture, 103
Chemin de Fer de Paris à Limours, 103
Chenu, Jeanne Marie, 125–26
Chevreul, Michel, 47–48, 50–51, 95
child predators, 119–20
children, 54–58, 70–74, 112–20
cholera epidemic, 38–39
City Beautiful Movement (U.S.), 157
class, 69, 90, 132, 137, 141–42
climate in Paris, 24
communitas, 7–8, 94, 106, 128, 131, 152
community relations. *See* neighborhood and community relations
complaints. *See* concessions and concessionaires; neighborhood and community relations
Comte, Auguste, 5, 44
concerts, 107–9
concessions and concessionaires: children and, 116–17; for cricket, 147; hirings and firings, control over, 185n38; for ice skating, 142–43; petition process, 77; primary interest of, 61; rent forgiveness and flexibility with, 82–84; rules and restrictions on, 78–82; small concessions as social safety net, 84–85; structures, concerns about, 80–82; types of, 77–78; women concessionaires and *receveuses*, 85–91, 109–10
Congrès Internationaux d'Architecture Moderne (CIAM), 158
Congress of Vienna (1815), 14
Corbin, Alain, 127
cricket, 146–48, 192n63
criminals and crime, 121–22
cycling, 148–51

Dabot, Henri, 119, 138
Darcel, Jean, 53, 65–67, 83, 174n44
Davioud, Gabriel, 96, 136

Delvau, Alfred, 130, 133
depopulation concerns, 54, 57, 58, 114
Descieux, Louis Cyprien, 55–56, 73
design and layout alteration petitions, 111–15
Donné, Alfred, 54–55
Dubois, Pierre Charles, 144
Du Breuil, Alphonse, 22, 47
Du Camp, Maxime, 34
ducks, 25–26
Dupeche (guard), 119

École Communale des Garçons, 73–74
École municipale et département d'horticulture, 22
Eliot, Charles, 156
employees. *See* workers in greenspaces
entrances to parks, petitions on, 95–106
existential *communitas*, 7, 94, 128
Exposition Universelle (1867), 33–35

Faucheux, M., 88–89
Felten, Mme., 88–89
fêtes foraines (carnivals), 106–7
firewood gathering regulations, 122–23
fish, 27
fishing and fishing licenses, 27, 143–46, 191n49
Flachat, Stéphane, 39
Flamel, Nicolas, 17
Flammant, Louis, 76
Fleuriste de la Muette, 21
Fonssagrives, Jean-Baptiste, 37, 50–51, 52, 56–57
Forestier, Jean-Claude Nicolas, 156
Forney, John W., 133
Fournel, Victor, 32–33, 171n82
Frégier, Henri-Antoine, 39–40, 46–47, 52
gangs, 120–21

Garde Républicaine, 106
Gérard, M., 25–26
Gilles, M., 144
Glacière station, 103, 111
gloire of France, 28, 57, 153
Goudon, Édouard, 135

Grad (guard), 72
Grand Chalet, Bois de Boulogne, 136
Gras, M., 120–21
grassy areas, 23–24
Grébauval, Armand, 105, 110
greenhouses, 21
guards: children and parents, interactions with, 70–74; conduct regulations and public interactions, 65–70; creation of Garde du Service, 62; functions of, 61; housing for, 64; interactions with other employees, 64–65; military model for, 62–63; nature of employment as, 187n77; Pissot's admonishment of, 123; public indecency incidents and, 119–20
guingettes de barrière, 147, 191n58

habitants, 95. *See also* neighborhood and community relations
Haussmann, Georges-Eugène: Alphand compared to, 19–20; appointment as prefect, 1, 15–16; in Bordeaux, 174n44; criticism of, 32; horse and carriage rental rules of, 78; modern city, view of, 28; park service and, 18; public health and, 52–53; sports and leisure and, 143, 146–47; studies on, 2
health, public: Alphand and, 53–54; children's health, fresh air, and exercise, 54–58; cholera epidemic and following reformers, 38–47; fountain removal at Square Montholon and, 112–15; *Le Monde illustré* article on, 16; Napoleon III and, 37, 52–53; positivism and, 37–38, 43, 44, 48; property rights vs. public interests and, 44–45; trees and plantings, value of, 47–52; utopian socialism and, 37, 39
Hiolle, Ernest-Eugène, 118
Hippodrome of Longchamps, 134–35
Hirtz, Dr., 87
homeless, 122–25
homosexual activity, 67, 177n26
horse races, 134–35
horticultural museum proposal, 28–29

horticulture students, 21–22
hours of parks, 102–3
Houssaye, Arsène, 57, 136
Hufeland, Christoph, 54, 56–57
Hunt, Lynn Avery, 166n15

ice production, 27–28
ice skating and ice safety, 135–43
Île de la Grande Jatte, 130–31, 159
Innocents, Square des, 122
international landscape architecture, 155–59
Invalides, Square des, 67–69

jardin, defined, x. *See also specific gardens by name*
jardin anglais style, 8–9, 20, 133, 153–54, 166n15
jardin français style, 20
Javelot, M., 79–80
Jeannel, Jean-François, 50–52, 174n44
Jessaint, Square, 71
Jockey Club, 134–35
Jouanet, Alexandre, 17–18, 22, 35, 50, 51–52
Jouas, Jean, 76

Kahn, M., 145
Kempfen, M., 147–48
kiosques de musique, 107–9
Kohler (guard), 122

LaBrousse, M., 148
Lacoste, Mme., 87
lakes and *tour des lacs*, 132–34
Lamartine, Square, 121
landscape architecture, international, 155–59
landscape crews. *See* cantonniers
Larcher, Mme. Amélie, 126–27
Laurens, Mme., 81
lawns and grassy areas, 23–24
layout and design alteration petitions, 111–15
Liebig, Justus von, 49–50
Le Nôtre, André, 8, 20, 57, 154
Lévy, Michel, 40–43, 47, 54–55, 190n40
Lhuiller, M., 121

liberties and rights, 6–7
lighting, 103, 105
Lion, Inspector, 110, 150, 151
Littré, Émile, 5, 44
Loliée, Frédéric, 134
London, 9, 33–34
Louis XVI, Square, 87, 95–96, 97(map), 103
Luxembourg, Jardin du: alteration of, 32, 171n82, 183n14; Delvau on, 130; *gloire* of France and, 57; maintenance complaints, 109; Medici Fountain statuary scandal, 119; park service jurisdiction and, 53
Lyons, 29

Mairie, Square de la, 118
maisons de tolérance (brothels), 125–26, 127
Marceaux, M., 109
Mare St. James, Bois de Boulogne, 139
memorialization, 17
Millet (guard), 70–72
Monceau, Parc, 32, 78, 177n26
Monge, Square, 69–70
Monsus, M., 81–82
Montholon, Square, 72, 86, 112–15, 113(map), 114(map), 116, 124–25
Montsouris, Parc: complaints about, 111; concerts in, 106; as fish hatchery, 27; fishing in, 145–46; guards in, 64; location of, 53; redesign petition, 115; Sceaux station and access issues, 103–5, 104(map); Third Republic and, 53
Moreaux (guard), 68–69
Morel, Jean-Marie, 166n16
Mulat, M., 138
musical performances, 107–9
mutual aid societies, 75, 107, 141

Napoleon III: city renovation and, 1–2, 14–15, 35; "The Extinction of Pauperism," 37; Haussmann and, 16; involvement in parks, 100, 183n14; mutual aid societies and, 75; petition to, 100; positivism and, 5; public health and, 37, 52–53

nationalist sentiment, 28–29, 35, 57, 153
neighborhood and community relations: children and, 115–20; *communitas* and, 94, 106, 128; community and social events, 106–7; criminals and crime, 121–22; entrances, access issues, and traffic flow, 95–106; the indigent, vagabonds, and homeless, 122–25; layout and design alteration petitions, 111–15; maintenance and management complaints, 109–11; musical performances, 107–9; prostitutes, 125–27; and rights, sense of, 128–29; *riverains* and *habitants*, 95; youth gangs, 120–21
New York City, 9, 156, 158
night patrols, 123
normative *communitas*, 7, 94
nurseries, 21

Observatoire de Paris, 153
Olaqnier, M., 31–32
Olmsted, Frederick Law, 156
Ottin, Auguste, 28–29, 119

parc, defined, x. *See also specific parks by name*
Parent-Duchâtelet, Alexandre-Jean-Baptiste, 39–40
Paris. *See specific places and topics*
Paris Cricket-Club, 146–47
Paris Guide, 34–35
Parmentier, Square (now Square Maurice Gardette), 53, 101–2, 101(map), 106
Payes, Louise, 125–26
Pelletan, Eugène, 34–35
Peniston, William, 67
Persigny, Duc de, 14, 15
petition campaigns. *See* neighborhood and community relations
Pfeiffer, M., 151
Pichot, A., 93
Pinkney, David, 2
Pissot, Auguste: duck problem and, 25–26; guards and, 60, 64, 123, 187n77; regulations and, 137–38, 144, 148; report on cricket,
146–47; report on ice skating, 139, 140–41
place, defined, x
place (*lieu*) and space (*espace*), 7, 156–57
Place de La Réunion, Square de la, 53
Plantes, Jardin des, 53
pleasure gardens, 130
population growth, 130
positivism, 5, 37–38, 43, 44, 48
promenade, defined, x
property rights, private, 44–45
prostitutes, 125–27
public and private, concepts of, 5–6, 44–45, 137

receveuses, 85–87, 90, 109–10
rent forgiveness for concessionaires, 82–84
restaurant concessions, 78–80
Richard, M., 149–50
rights, 6–7, 128–29
riverains, 95, 112–13. *See also* neighborhood and community relations
Robert, Daniel, 119
Roberts, Henry, 45–46
Robinson, William, 1, 57–58
Roche, M., 89
Rouen, 30
Rousseau, Jean-Jacques, 166n16

safety issues, 103, 138–39, 148, 149–50. *See also* guards
Saint-Simon, Henri de, 5
Saint-Simonian movement, 37, 39
Saussure, Nicolas-Théodore de, 49–50
Say, Léon, 142
Sceaux station, 103–5, 104(map)
Seilheimer, M., 102–3
Selves, Justin de, 151
Service des Promenades et Plantations: Alphand as head of, 18–19; citizens, interplay with, 1–2; criticism of, 32; formation and structure of, 18–19; Garde du Service, creation of, 62; internal culture of, 155; international landscape architecture, influence on, 155–59; multiple use of

Service des Promenades et Plantations (*continued*): greenspaces and, 131; records of, 4. *See also specific persons, parks, and topics*
Seurat, Georges, 95, 130–31, 159
Siedlungen movement, Germany, 157
Simard, François, 119
social welfare, 84–85
Société de médecine publique et de génie sanitaire, 153
Société de Secours Mutuels dite du Boulogne, 75
Société laïque d'appui fraternel, 107
soil conditions, 22–24
Solidarists, 151
spatial practices, de Certeau on, 7, 156–57
sports and leisure activities: *communitas* and, 131, 152; cricket, 146–48, 192n63; cycling, 148–51; fishing, 27, 143–46, 191n49; horse racing, 134–35; ice skating, 135–43; lakes and *tour des lacs*, 132–34; multiplicity of use and conflicts over, 131–32; in Seurat's *Grande Jatte*, 130–31, 159; urbanization and, 130
square, as term, x. *See also specific squares by name*
students of horticulture, 21–22
Suard, M., 120–21
Sumpter, Thomas, 146

Tardieu, Antoine Ambroise, 43–45
Temple, Square du, 16, 17, 23, 58
Thays, Jules Charles, 156
thieves, 121–22
Thouin, Gabriel, 133
Tiger, M., 115
Tisset, M., 122
tour des lacs, 132–34
Touring-Club de France, 151

Tour Saint Jacques, Square de la, 17
traffic flow issues, 95–106
train stations, 103–5, 104(map), 111
Travet, M., 145–46
trees, 21, 22–23, 47–52, 79–81
Trouble, Mme., 87
Tuan, Yi-Fu, 3
Tuileries, Jardin des, 53, 57
Turner, Victor, 7–8, 94, 152

urbanization, 130
utopian socialism, 37, 39

vagabonds, 123–24
Vaquier, M., 122
Varé, Louis, 132
Véron, Mme., 110
Ville, Georges, 50
Villermé, Louis, 39–40, 54
Vintimille, Square de, 62
Vosges, Place des, 53, 73, 85, 87–89, 93, 106–7, 120–21

wildlife, 25–27
women, 69, 85–91
workers in greenspaces: cantonniers, 64–65, 74–76, 110; categories of, 61; concessionaires, 61, 76–92, 109–10; guards, 61, 62–74; horticulture students, 21–22; ice-rescue team, 139; social realm and, 60–61

yacht racing, 148–49
youth gangs, 120–21

Zubler, M., 144